PHOTOVOLTAICS
Sunlight to Electricity in One Step

PHOTOVOLTAICS
Sunlight to Electricity in One Step

PAUL D. MAYCOCK
and
EDWARD N. STIREWALT

Brick House Publishing Co.
Andover, Massachusetts

To my wife

Roma L. Maycock
"God willing, a priest"
—P.M.

To my mother
Evelyn Fraser Stirewalt
—E.S.

Published by **Brick House Publishing Co., Inc.**
34 Essex Street
Andover, Massachusetts

Production credits
Editor: Jack Howell
Cover design: Bob McCormack
Interior design: Herbert Caswell
Illustrations: Michael Prendergast
Copy editing: Elizabeth Holland
Rewriting: Lloyd Corwin
Typesetting: Neil W. Kelley
Production supervision: Dixie Clark Production

Printed in the United States of America

Library of Congress Cataloging in Publication Data

Maycock, Paul D.
Photovoltaics, sunlight to electricity in one step.

Bibliography: p.
Includes index.
1. Photovoltaic power generation.
I. Stirewalt, Edward N. II. Title.
TK2960.M39 621.31'244 81-6157
ISBN 0-931790-24-7 AACR2
ISBN 0-931790-17-4 (pbk.)

Contents

CREDITS

The illustrations in this book are used through the courtesy and with the permission of the following sources:

Electric Power Research Institute: Fig 3–15; *Martin Marietta Aerospace:* Fig. 1–6; *Don Monroe:* Fig. 1–8; *N.A.S.A. Lewis Research Center:* Figs. 1–2 and 5–3; *Sanyo Electric, Inc.:* Fig. 1–3; *Solar Design Associates, Architects and Engineers,* Lincoln, Mass.: Fig. 1–4 (module installation photo by MIT), 1–7 (photo by Mobil Oil Corp., Inc.; *Solarex Corp.:* Fig. 1–5; *Statistical Abstract of the U.S. 1980:* Figs. 5–5 and 5–6; *U.S. Army photos:* Figs. 4–7 and 5–4; *U.S. Department of Defense:* Figs. 4–3 and 4–4; *U.S. Department of Energy:* Figs. 1–1, 2–5, 2–8, 3–1 through 3–4, 3–5 (with Westinghouse, Inc.), 3–6, 3–7 (with Crystal Systems, Inc.), 3–8 through 3–14, 3–16, 4–5 (with Southern Railway), 4–6, 4–8 through 4–12, 5–1, 5–2, 5–6, 5–7 (with Aerospace Corp.), 7–1 (with Garrett, Inc.), 10–1 through 10–6.

Preface

In the Fall of 1979, we set out to write a relatively simple book about photovoltaics that would explain in a straightforward way what this technology is about, how solar cells work, and how solar photovoltaic systems may help us with our energy problems. A colleague, Andrew Krantz, first pointed out the need for a book written for the general public, to fill the void between government documents, pamphlets, and studies, on the one hand, and college-level textbooks on the other.

What began as a simple task turned into an odyssey of sorts. It soon became apparent that we were dealing with a potentially potent force in society, the impact of which we could but faintly discern. We became ever more deeply convinced that photovoltaics could have wide ranging effects on some of today's most pressing and perplexing problems—pollution, inflation, key industries now in turmoil such as the automobile, electric utility, and housing—and on the yearnings of individuals and of nations for greater energy independence. The trail at times led where we least expected it might.

Many people have helped. After the first draft was completed, Dr. Lloyd Corwin assisted in guiding us through a difficult transition stage, rearranging sections and rewriting most of the original text. The final document reflects his organizational competence and his smooth writing style. We gratefully acknowledge his able editorial assistance and professional counsel.

James N. Stirewalt, young economist and historian, after digging through voluminous market development studies on photovoltaics, had a principal hand in bringing Chapter 5 into focus. He also contributed the sections on the history of photovoltaics that appear in Chapter 4, and an appendix. In addition, Jim organized our reference files, indexing some 300 original documents and citations, without which the work could never have been completed. To one of us particularly, his occasional calming words provided reassurance when the road ahead seemed at times endless. We are deeply indebted to him for his devotion to the project and his steadfast determination to see it through.

Dr. Morton B. Prince, a research scientist who has made noteworthy original contributions to the technology, was especially helpful on several points and allowed us occasional access to his personal library. Our special thanks are extended to him for his helpful support.

To Elizabeth Holland, assistant editor of *Solar Age* magazine, goes highest praise for a superb job of editing. Her deft pen pruned away the dead wood to clarify every sentence, while managing never to do violence to the basic style of the writer. We are grateful for her perseverance with a complex manuscript and the inevitable changes associated with its completion.

We were fortunate to have a publisher who was willing to work with us over a period of months toward a common objective. We are especially indebted to Jack D. Howell and James L. Bright of Brick House Publishing Company for their patience, their support and their invaluable advice.

The major typing burden was carried at successive stages by Cheryl Schuler and by Maggie Ruiz, and we are indebted for their diligent support. We also wish to thank Elizabeth Bennett and Marcia Stirewalt for earlier typing assistance.

We are indebted to many others in various laboratories and industrial firms for the use of photographs and other illustrations, which were generously provided. Our thanks to Tim Taverner for reviewing portions of the manuscript and providing reactions from the perspective of one whose hands are daily in the silicon cell development process.

Our respective employers were considerate in giving us a free hand to address the subject as we saw fit. Not once did anyone—superior or colleague—ask to look over our shoulders. We thank them for the independence afforded us. We reciprocated by striving and essentially succeeding in keeping this effort completely out of our respective offices. Weekends, nights, and holidays were consumed over a period of sixteen months, and manuscripts accompanied us even on vacations, as our families know too well.

So to our wives especially, who shared and endured when all spare time seemed devoted to a book, we offer a full measure of gratitude for their patience, understanding and constant support.

We come away sensing incompleteness, knowing that what we have recorded is just the beginning of a fascinating saga. For what has been left undone, for what may be amiss, for the shortcomings as well as the major conclusions to which we were led, we alone take full responsibility. In the end, we have realized that the issues of photovoltaic development are not really our problem, but a central piece in America's own energy odyssey. We must leave it to you, O Reader, which we now gladly do, confident that the people will put photovoltaic energy to good use. Read on!

P.D.M.
E.N.S.

Fairfax County, Virginia
March 1981

Prologue

Just as prehistoric man fashioned simple tools from sticks and stones, so the post-industrial society, as John Kenneth Galbraith has characterized our age, picks up whole new technologies and puts them to use. In this era of high technology, a steady succession of new developments and capabilities comes at us. Since World War II gave birth to radar, jet planes and nuclear weapons, dozens of new products have appeared: television, space ships, satellite communications, electronic computers, lasers, to name a few. Bioengineering is now the rage on Wall Street.

The average person has had precious little say about which technologies should be developed, or why. As Galbraith has pointed out, these decisions are made for us in the corporate boardrooms and the centers of finance. Indeed, when a new product is foreseen, its market is developed simultaneously with the product. The desire to buy is instilled in the public before the first unit appears on a retail counter.

Occasionally the public, or some element of it, has rebelled openly at a particular trend: the spread of nuclear reactors, the development of a supersonic transport, the misuse of some forms of telecommunications that invade privacy. Such rebellion is rare; usually people simply enjoy as they will, or tolerate, the fruits of science and engineering and American ingenuity at commercializing them.

The mark of a mature post-industrial society is the ability to choose intelligently among new technologies as they present themselves, selecting those that meet present needs and long term goals. The special challenge now is to match together two or more technologies so that they augment or complement each other for some desirable purpose.

Energy technologies are now the focus of concern. Energy—its sources and uses—is central to everything from the economy to the ecology, from domestic social concerns to international relations. In the area of energy, as never before, wise choices are needed. Decisions made today can commit us for years to a particular direction.

This book tells the story of one new energy technology—how it began and how it works, its current status, and its potential. Fortunately, with photovoltaics, the average person can yet have a major hand in choosing what road we shall follow. For this fleeting moment, the future use of an energy source is what the people will make of it.

CHAPTER 1
Setting

Photovoltaic* devices—the solar cells which convert sunshine directly into electric current—first came to the attention of the general public with the advent of America's space program. Like the vehicles which carried them, photovoltaic cells were seen as exotic, impractical, and expensive. The solar cells that accompanied the Vanguard I were painstakingly made by hand, and cost more than $1,000 per Watt. Such a cost would be clearly prohibitive for a typical household, which needs 5,000–10,000 Watts to satisfy its peak electrical demand.

In nearly a century and a half since the photovoltaic effect was first observed in 1839, the technology has grown from a laboratory curiosity into a mature science. With the myriad practical uses found for it in the last twenty years, photovoltaics is on the brink of exploding into the energy marketplace.

Originally hand-built and labor-intensive, solar cells can now be made on an assembly line basis. Originally designed for the most specialized of uses, they can now be found on self-powered wristwatches, navigation buoys, and buildings. Prohibitively expensive at first, they are now at most only a very few years from being directly competitive, on a cost-per-kilowatt-hour basis, with electricity generated by traditional water-, coal-, oil-, or nuclear-driven plants.

Photovoltaic solar electric power, contrary to the common perception, is no longer an exotic technology. To give some idea of just how far the reality of photovoltaics departs from the popular misconception, here are a few predictions:

1. Photovoltaics will be fully economic for massive private use before a major utility can design, purchase, and install its next new nuclear reactor.

---

* Pronounced fō′-tō-vōl-tā′-ĭk.

Fig. 1–1 Daytime WBNO, a radio station broadcasting from Bryan, Ohio, draws most of its power from a 15-kW peak photovoltaic array. A battery is used to store excess energy.

2. Photovoltaic systems installed on the roofs of residences in the United States will be fully economic—delivering electricity that costs five to ten cents per kilowatt hour—by 1986 without tax rebates, and by 1984 with a 40 percent tax rebate and Solar Bank financing.[1]
3. If we seriously begin to adopt photovoltaics now, as much as 30 percent of the nation's electric energy can come from this source by the year 2000. This is roughly three times the market penetration of the nuclear option today.

Photovoltaics offers the quickest and surest way out of the growing energy crisis in which we are mired. We must treat this technology as a practical reality to be integrated both into decisions about energy policy

[1] The 40 percent tax credit for users of renewable energy sources, including photovoltaics, was enacted into law in the summer of 1980 by the Crude Oil Windfall Profit Tax Act of 1980, Public Law 96-223. Section 202 amends 26 U.S.C. 44C (b)(2) to provide 40 percent tax credit on the first $10,000 of residential renewable energy source expenditures, i.e., to a maximum of $4,000. See *Statutes at Large,* Vol. 94 (1980) and "Solar Energy Tax Credits Pegged at 40 Percent," *Solar Engineering* 5, March 1980, p. 9.

and supplies, and into the technical and political assumptions that underlie our economy and sustain our society.

In a very real sense, and in whatever form, energy is the lifeblood of this society. Whatever the source—oil, coal, hydro, nuclear, gas—the use of energy is all-pervasive. Energy lights and heats our homes, offices, and factories. It powers the machines of industry and transportation. The clothes we wear, the food we eat, the buildings where we live and work, the goods we produce, such as dye, lubricants, metals, glass, plastic, and the services we supply, such as transportation and communication—energy is inextricably woven into it all.

There will be no better time than now to face the reality that our store of fossil fuels, our hydro sites, even our supplies of uranium, are finite. The OPEC nations, unwittingly and for reasons other than altruism, have done the world a favor by bringing to our attention, in stark and unambiguous terms, this central fact.

We in the United States are one-twentieth of the world's people, yet we consume one-third of its energy. Ignoring questions of equity for the moment, the interruption of energy supply is more disruptive to a highly industrialized society such as ours than to a less-developed one. Since the first oil embargo of 1973, we have recognized our vulnerability. In spite of hand wringing and lip service, we have made little genuine progress toward resolving our energy dilemma.

In the absence of a focused national commitment, the real and perceived energy shortages have exacerbated tensions between different elements of our society. The widespread conviction that we are being taken for a ride by "big oil," and the bumper stickers that read "Drive Eighty and Freeze a Yankee" or "If you like our Post Office, you'll love nationalized oil" are symptomatic. Another consequence of our energy situation is the chronic and massive balance-of-trade deficit. It is small comfort that other industrialized nations lacking substantial domestic supplies are economically hurt more than we are.

This trade deficit aggravates the inflationary spiral begun during the Vietnam war years. At the same time that rising oil prices increase the costs of heating, manufacture, and transportation, the billions of dollars we send the OPEC nations exert a devaluating influence on our currency. An editorial in the *New York Times* points out that "energy is still the problem—inflation is merely the mirror . . . no one is going to solve the inflation problem without first solving the energy problem."[2] President Carter said that nothing he ever confronted in his career is comparable in complexity and difficulty to the energy problem.[3] Belatedly, six years

[2] "Energy is Still the Problem," *New York Times,* 29 July 1979, p. 18-E.
[3] Press conference, 19 October 1979.

after the first embargo, a President explicitly confirmed that the Cassandra warnings about our heavy reliance on foreign oil were correct. He called it "a clear and present danger."[4] To the considerable extent to which other fossil and nuclear fuels are interchangeable for generating electricity, the warning can be generalized: we *must* find alternatives to our current fuels. It is obvious that something must be done to stanch this economic hemorrhage.

On the one hand, we long for ultimate solutions to the energy problem. The breeder reactor, some argue, would save us if only environmentalists and kindred malcontents would get out of the way. The fusion generator will provide us with cheap energy, but it may require that we invest billions and wait 50 years. In another scheme, orbiting photovoltaic satellites could beam energy to earth by microwave where it would be reconverted to electricity in quantities to meet the total U. S. demand.

On the other hand, such near-term solutions as synfuels, more nuclear plants, coal-to-methane conversion, and others, are at root a recommendation for the "more of the same" attitudes that got us in trouble in the first place. They commit us even further to the existing high-capital, central-distribution model that involves increasingly unacceptable external costs: pollution, containment of radioactive wastes, supply monopoly, operating and maintenance costs, and inelastic fuel supply.

More of the same, in addition to incurring these external costs, may create new problems. For example, in the summer of 1979, Congress and President Carter were pressing ahead with a massive synthetic fuels program, which involves mining and processing vast reserves of coal and oil-rich shale in several western states to produce a number of synthetic fuels (gas, liquid fuels, and refined coal, among others). However, processing these deposits requires large amounts of water, a precious natural resource perennially in short supply there. By October of that year, the President found it necessary to make a special trip to a conference of western governors to promise them, in person, that their water resources would not be sacrificed to an otherwise urgent national synfuels program.[5]

We live in a highly technological society, and an inextricable element of this is our massive need for energy. This need is neither good nor bad in

---

[4] State of the Union Address, 23 January, 1980. President Carter said, "The crises in Iran and Afghanistan have dramatized a very important lesson: Our dependence on foreign oil is a clear and present danger to our national security." In the same speech, he indicated that the United States was willing to use military force to protect "vital interests" in the Persian Gulf and called for renewal of draft registration.

[5] When the synfuels program was subsequently approved by Congress in the summer of 1980, a key feature of the administration's energy program was lost: the creation of an Energy Mobilization Board with authority to expedite critical energy projects by setting aside normal review procedures and controls by federal agencies and state and local governments.

Fig. 1–2 The photovoltaic-powered RAMOS weather station at the edge of New Mexico's vast eastern plains reports weather data hourly via satellite to National Weather Service offices.

and of itself. We have devoted enormous amounts of capital to provide for our energy needs, and patterns of decision-making about energy have been institutionalized to the point that it is now very difficult to think about energy in other than "more of the same" terms.

This "more of the same" attitude tries to treat the energy crisis as a supply problem—solvable with more mines, more wells, and more reactors—when in reality it has become a structural problem. For the most part, this approach is an extrapolation of the energy supply strategies we have followed since centrally produced electric power first became a reality. Considering the number of people, physical plants, and equipment already committed to this structure, such a response is understandable, but it is wrong-headed.

There is no need to rehearse here the shortcomings of the centralized fossil fuel and nuclear model to which our country and all industrialized nations are presently committed. By the same token, there is no need to do a cost-benefit analysis of the indirect methods of tapping the sun: windpower, driven by the sun's heating the earth and its atmosphere; ocean thermal, which exploits the sea's temperature differentials; biomass, which binds the sun's energy into chemical bonds in plants. They should be explored; diversity of sources, rather than some single monolithic solution, is the best strategy. But this book is not a global discussion of energy policy. It is about photovoltaics and how this technology not only can help us out of our present plight, but also substantially change our world for the better.

We restrict ourselves when we think of solar power only as low-grade energy available as heat, which buildings can capture and distribute passively through proper location, orientation, and design, or collectors can capture and distribute "actively" with circulating liquids or forced-air systems. Several proposals for collecting sunlight, a very diffuse source of energy, involve concentrating its rays so that the resultant heat performs work such as generating steam, which in turn drives turbines to produce electricity. These systems commit us to the fixed, capital-intensive, cumbersome, complicated central-distribution model that is a part of our current energy problems.

In contrast, photovoltaic technology is both simpler and more sophisticated than either the traditional fuel-driven or the proposed solar thermal-electric plants. Rather than elaborate and costly machinery—all too subject to Murphy's Law—photovoltaics uses the fundamental properties of matter itself to produce power quietly, safely, and with virtually no operation and maintenance costs. Because photovoltaic systems are modular and can be produced in size and type to fit almost any need, they permit local, decentralized installation at the point of need, and can,

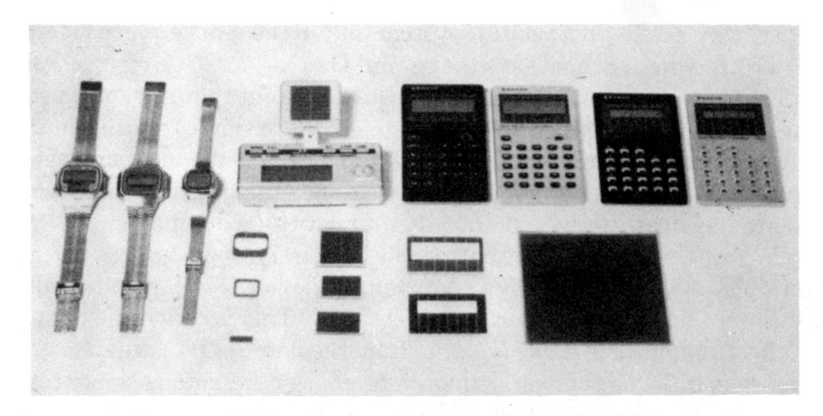

Fig. 1–3 These products manufactured by Sanyo use amorphous silicon photovoltaic cells.

in very simple fashion, be designed to match the output to the user's purposes.

The best recent analysis of our energy situation appears in the Harvard Business School study *Energy Future*.[6] With deletions, but without distortion, the authors' conclusions can be boiled down to this:

> The purpose of U.S. energy policy should be the managing of a *transition* from a world of cheap imported oil to a more balanced system of energy sources . . . The nation has only two major alternatives for the rest of this century—to import more oil or to accelerate the development of conservation and solar energy . . . Because environmental and health problems have increasingly become issues in the political process, the outlook for coal is considerably less promising than officially projected . . . Nuclear energy has a set of external costs even more controversial [than coal] . . . At most, during the next dozen years nuclear energy might add a million and a half barrels a day of oil equivalent thus reaching three million barrels per day. This level would require the continued operation of *all* existing nuclear capacity plus the completion and operation of all new capacity currently under construction or on order. That we stress is a very bullish scenario. . . . The cornerstone of our thinking is that conservation and solar energy should be given a fair chance in the market system to compete with oil and other traditional sources . . . a more significant change could occur from the widespread use of on-site solar technologies.

What then is the problem? If photovoltaics is safe, clean, and reliable, and if there has been a substantial reappraisal of the role of energy in our

[6] Robert Stobaugh and Daniel Yergin, eds. *Energy Future: Report of the Energy Project at the Harvard Business School* (New York: Random House, 1979).

society, why aren't photovoltaic systems out in the marketplace where the Harvard Business School says they should be?

The answer is that photovoltaic systems are out there for some uses, in some places. Part of the problem is that people are slow to give up outdated ideas about photovoltaics which match the reality of twenty years ago, not of today. One such idea, which happens to be wrong, is that photovoltaic cells are physically and economically impractical to produce.

Part of the problem is that people generally don't know very much about photovoltaics. Briefly, photovoltaic devices—solar cells—absorb sunlight and convert it directly in electricity. This is called the photovoltaic effect; "photo" for light, "voltaic" from the name of the Italian scientist, Volta, who gave us the volt. When light energy or photons strike certain materials called semiconductors, internal voltages are created. The basic scientific principles which underlie this effect are well-understood. The solar energy available on a bright summer day at noon is about 1,000 Watts (1 kilowatt) per square meter. This amount of solar energy falling on a single square meter, if converted perfectly to electricity, could power ten 100-Watt bulbs, or two furnace motors, or several 25-inch television sets.

It is not, of course, high noon and summer all the time, and photovoltaic cells do not work at 100 percent efficiency. They presently work at about 15 percent efficiency, with the best performance being about 25 percent. And some areas of the country receive more light than others. But, the same misconception that has hampered the development of passive solar design, the assumption that it will only work in such locations as the Sunbelt or at certain times of year, has dogged photovoltaics.

It is true that in a year Boston receives only half as much sunlight as Phoenix. But the local cost of electricity is a variable of greater potential significance for determining the suitability of a photovoltaic installation. While sunshine varies by a factor of two in the United States, electricity rates vary by a factor of more than six. In late 1979 the Boston home owner paid $.0814 per kilowatt hour; his Phoenix counterpart paid $.0556. In New York City, where electricity cost $.139 at that time the potential market for photovoltaics may in fact be better than in Phoenix, even though New York receives less sunshine. Of course, there are places where it would be hard to make a credible argument for large-scale reliance on photovoltaics—in Seattle, for instance, where hydroelectric power cost only $.0222 per kilowatt hour in late 1979 and heavy clouds cover the sky for much of the winter.

Another part of the problem is that in our highly industrialized society it is impossible to change course suddenly and very difficult to do so even gradually. There is no deliberate conspiracy to keep photovoltaics out of the marketplace. But there is a lack of commitment to it, due in part to the

Fig. 1–4 Photovoltaic solar energy house. The Carlisle House in Boston, completed in Spring 1981, was specifically designed for a 7.5-kW peak photovoltaic array. House also incorporates passive solar design and a solar thermal hot water heater. Above: Installing photovoltaic panels on 45-degree sloping roof. Below: Architect's rendering of completed residence.

popular misconception that photovoltaics has little to offer in this century. Political leadership has done more than pay lip service to photovoltaics, but as yet has been unwilling to give it a high place in national priorities.

Any major change in something so basic as the way we produce and allocate increasingly dear energy supplies must have the broadest understanding and support of the people affected by those decisions: ourselves. It is only recently that solar power in general and photovoltaic technology in particular have had a credible and reasonably well-informed constituency. The Solar Lobby, a coalition of groups numbering over 80,000 persons, was formed in 1978 as a result of Sun Day.

Government had previously shown some interest in solar energy. As early as 1972, the National Science Foundation wrote that

> . . . a substantial development program can achieve the necessary technical and economic objectives by the year 2020. Then solar energy could economically provide up to (1) 35% of the total building heating and cooling loads; (2) 30% of the Nation's gaseous fuel; (3) 10% of the liquid fuel and (4) 20% of the electric energy requirement.[7]

Six years later, after unparalleled price increases for imported oil, the President's Council on Environmental Quality estimated that photovoltaics might provide 2 to 8 quads (one quad equals a quadrillion Btu's) of energy annually by the end of the century.[8] Photovoltaics is projected in the estimate to outstrip the nearest competitor, wind power, by as much as two and one-half times by the year 2020.

It should not go unnoticed that under three administrations, the budget for photovoltaics has risen steadily from $4 million in 1973 to $160* million in 1982. (Even so, the present level is only three-hundredths of one percent of the national budget, and about 10 percent of the nuclear research budget.) Congress has also supported the solar effort through significant special legislation. In 1978, the Solar Photovoltaic Energy Research, Development and Demonstration Act (Public Law 95-590) which sets as a goal the doubling of photovoltaic production annually for

---

[7] *An Assessment of Solar Energy as a National Energy Resource.* NSF/NASA Solar Energy Panel, December 1972, NSF RA-N-73-001 (Springfield, Va.: National Technical Information Service, 1973), p. 5. A year later, NSF sponsored a workshop conference at Cherry Hill, N.J. This 3-day meeting was the seminal event which launched the federal effort to make photovoltaics economically useful for terrestrial applications. See *Photovoltaic Conversion of Solar Energy for Terrestrial Applications,* Oct. 23–25, 1973, Vols. I and II, NSF-RA-N-74-013; Executive Report, NSF RA-N-74-073 (Springfield, Va.: National Technical Information Service, 1974).

[8] U.S. Executive Office of the President, Council on Environmental Quality, *Solar Energy: Progress and Promise* (Washington, D.C.: U.S. Government Printing Office, April 1978), p. 6.

*Note: The Reagan budget for FY 1982 reduces this to $63 million.

ten years and authorizes the expenditure of $1.5 billion over that period, was passed. The law foresees a cumulative installed capacity of 4 gigawatts by 1988, by then at system costs of $1 per Watt in 1978 dollars. (Remember, the solar cells of the sixties cost more than $1,000 per Watt.) Recognizing the potential for change inherent in photovoltaics, the law also establishes a special advisory committee with broad public representation to help assure that the public interests are met.

In June 1979 an inter-agency Domestic Policy Review committee, called for by President Carter and headed by then Energy Secretary James Schlesinger, set a goal for the year 2000: solar would contribute 19 quads, out of the 95 required, or 20 percent of to the nation's energy supply.[9] The Harvard Business School study also envisions 20 percent of our energy needs being met by solar, and explicitly points to photovoltaics:

> One can certainly say that photovoltaics thus far constitutes a success story . . . advances can come quickly, for as one specialist has put it, photovoltaics are "embedded in what has been one of the most fertile environments for innovation in the twentieth century—the semiconductor industry." As has been the case in that industry, increased volume will bring increased experience, improved processes, and reduced costs. Investments in research will produce fundamentally new production techniques that could lead to even lower costs.[10]

The study urges that federal purchases of photovoltaic equipment be used to drive production costs lower, and even suggests that funds earmarked for the Clinch River breeder reactor be used for that purpose.[11]

Over the past several years there has also occurred in some quarters a fundamental rearrangement of the assumptions and attitudes about the role which energy should play in our society. This reconception appears in various guises, sometimes described as "limits to growth," "small is beautiful," "soft energy paths," or "appropriate technology." The proponents hold in common a vision of a decentralized society in which all energy comes from renewable sources, especially those that are continually replenished by the sun.[12]

---

[9] President's message on Solar Energy, 20 June 1979, Fact Sheet, pp. 1, 3. Also see *Domestic Policy Review of Solar Energy.* U.S. DOE TID 28834 (Springfield, Va.: National Technical Information Service, 1979), p. 50.

[10] Stobaugh and Yergin, *Energy Future,* pp. 209–210.

[11] ". . . If only a part of the $2 billion of federal funds slated for the Clinch River breeder reactor were directed instead to photovoltaic purchases, the $1,000-per-peak-kilowatt price could be achievable very soon—compared to a $5,000-per-peak-kilowatt estimated cost at Clinch River."

[12] Nuclear fusion and breeder reactors, which are sometimes classified as renewable energy sources because they could for all practical purposes provide unlimited supplies, are not included in the popular concept.

As the cost of traditional fuels escalates, it becomes possible for reasonable people to propose energy sources and supply structures that only a few years ago would have been regarded as unrealistic and visionary. Windpower, ocean thermal differentials, and biomass are no longer considered "far out" potential sources of energy. It has become intellectually respectable to side with the late Ernest Schumacher, author of *Small is Beautiful*, that bigness in institutions, business, industry, or government is not only not betterness, but may in fact be inimical to many human values and aspirations. Granted, as Daniel Greenberg recently pointed out in the *Washington Post,* prudent people have been a bit put off by solar energy proposals because of "the vast freight of political and philosophical enthusiasms that some of its friends have loaded onto it."[13] We solar enthusiasts have sometimes gone out of our way to proclaim our disenchantment with some of the ways technology has manifested itself, an attitude which along with its opposite, uncritical technolatry, too often becomes ideological name-calling.

Recently, however, there has been a growing affirmative consensus that in fostering technical development we should give more care to selecting technical approaches and systems that are better matched to human needs and aspirations. The key word is appropriate. There is nothing intrinsically wrong with a technology that fits the needs of the times, and there is nothing intrinsically wrong with bigness in the right place, helping toward proper goals.

One popular formulation of these sentiments is set forth by Amory Lovins, British physicist, author, and long an advocate of appropriate technology. An ideal energy system, he suggests, should first consist of a set of diverse sources, each doing what it does best. Second, these energy sources must be simple and understandable from the user's point of view. Put a different way, our solutions to the energy problem should be our tools, not our masters. The test is not whether a solution involves high technology, but whether the average person can make practical use of it. In addition, the form of energy considered appropriate should be of the right quality, right for the purpose it serves, and properly scaled to the load. Lovins distinguishes among four kinds of energy:

- Low-grade heat for low-temperature processes;
- High-grade heat for processes requiring high temperatures;
- Mechanical energy providing torque for mechanical functions;
- Electricity for electronic and pure electrical demand.

---

[13] "Solar Power Without Granola," *Washington Post,* 2 May 1979, editorial.

Fig. 1–5 The world's first solar-powered gasoline station in West Chicago, Illinois, is a demonstration by the Amoco Oil Company, Solarex Corporation, and Semix, Inc. The 5-kW array and battery bank began operating in July 1980. (Courtesy of Solarex Corp.)

By carefully choosing technology appropriate to its use, he argues, we can maintain our present standard of living while using half the energy we now consume. For example, electricity, which is high quality energy, should not be used for heating. Moreover, the by-product heat from electric generating plants should not be wasted, but put to practical use. The use of solar thermal energy for heating rates high on Lovins' list of appropriate function matched to energy quality. He has proposed that his "soft path" would preclude the need to add more conventional electricity generating capacity in the United States. It is unlikely that, as a practical matter, this is the case. Still, photovoltaics meets the "soft path" criteria proposed, and is an attractive source of energy in a number of important respects.

In one light, this is pretty heady stuff. But, somehow, there has remained an air of unreality about these activities. The net effect of the

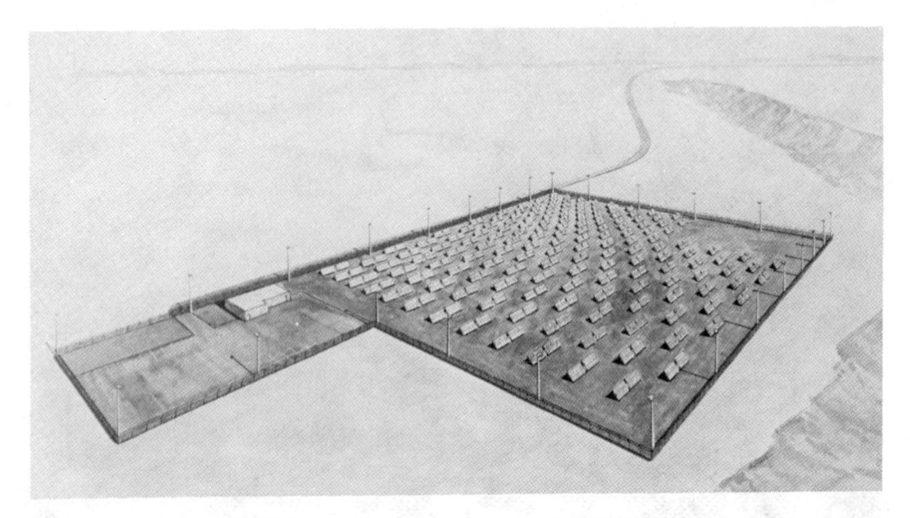

Fig. 1–6 In this 350-kW village power project being built in Saudi Arabia, a 2-axis tracking array produced by Martin Marietta is used. Each cell is covered by a plastic Fresnel lens.

projections by blue ribbon committees, academics, and some officials has been to establish in the public mind an impression that renewable energy, and specifically photovoltaics, is meant for the next generation, and certainly offers no possible help in the present energy crisis. Many of these studies are shot through with a presumption of helplessness. Like the use of outdated energy supply assumptions and strategies for projections into the future, the idea that photovoltaic technology is only for the next century is simply wrong, and needlessly impedes present-day commitments to photovoltaic development.

In another way, simplistic solutions to the energy problem, often presented by some quarters of the solar movement, unfortunately have contributed to the same air of unreality and heightened the public sense of frustration. Solar energy does not mean an across-the-board return to old-fashioned woodstoves and windmills. Far from it. But we do not belittle what the solar enthusiasts have accomplished. Their passion is a reflection of a deeper yearning throughout our society for a new day in energy.

The rapidly developing photovoltaic technology is being whipsawed in these tides of debate while its solid technical base, now well established and growing, is ignored. There is no need to resort to "technological timidity in a country that is teeming with technological strength," as Daniel Greenberg puts it.[14] The extensive development effort now under way, from basic research to engineering design—including rapid innova-

[14] Daniel Greenberg, "Technological Timidity," *Washington Post,* 22 May 1979, p. A-15.

Fig. 1–7 Photovoltaic Desalination Unit now in operation in Saudi Arabia. This 8-kilowatt unit pumps water from a shallow brackish well and desalts it by the reverse osmosis method.

tion in manufacturing techniques and full-scale tests of systems in the real world—virtually assures that inexpensive photovoltaic systems with conversion efficiencies approaching 20 percent will be on the market by the mid-1980s. In many areas and for special uses, large-scale first-generation production units can be in place by 1982, and these first-generation photovoltaic systems will become economically competitive for residential use in some parts of the country as early as 1984.*

Photovoltaics is a rapidly evolving technology. New cell design concepts, new materials from which to make cells, steadily increased efficiencies, and simplified production techniques virtually assure that the mass production of low-cost, long-lived cells is no more than two or three years away.

Consider the following scenario:

- Costs will continue to decline, just as they have historically, with a pronounced drop occurring when automated factories begin high-volume production of thin film or ribbon arrays.
- Residential uses will grow at an exponential rate, and in the latter half of this decade utilities and their customers will work out new relationships regarding the production, use, and pricing of electricity (an area in which there are already some interesting precedents). Simultaneously, the utilities will begin to experiment with their own photovoltaic collectors as these become relatively more attractive

* If the Reagan budget cut prevails, this date will slip at least two years, to 1986.

Fig. 1–8 The Solar Challenger, a photovoltaic-powered plane, had its first flight on November 20, 1980. Paul J. MacCready, president of Aerovironment, Inc. of Pasadena, Calif., constructed the plane to heighten peoples' awareness of the potential of photovoltaic cells. The craft has made 60 flights: longest distance, 15 miles; longest time, 1 hour, 55 minutes; highest altitude, 3,000 feet. A Paris-to-London flight will be attempted in June 1981.

when measured against fossil and nuclear plants which carry high capital, fuel, and operation and maintenance costs.

- Homeowners will begin to experiment with energy storage systems to see them through the evening hours and the inevitable rainy days. (Here again, there are imaginative technological developments, including improved batteries and inertial storage in flywheels.)
- In addition to supplying electrical power for household uses, including air conditioning and space heating (normally by heat pump), photovoltaic systems will gain increasing popularity as the principal source of power for the family automobile. One thousand square feet of photovoltaic array, costing about $7,000 in 1986, will take care of household power needs, except for air conditioning and heating. Double this array, and photovoltaics can recharge the battery for powering the family auto.

- Banks or financial institutions will be quite happy to lend money for a system guaranteed to last twenty or more years. Household insurance will routinely include wind, hail, or storm damage to the array.
- The customer can service his own system, obtain service as needed, or buy an annual maintenance contract.

By 1990, photovoltaic systems can be the preferred, most economic energy option for providing electricity[15] for homes, offices, schools, automobiles, and light industry, as well as for remote sites, miles from any central utility, throughout the United States and major parts of the world.

How we proceed to develop photovoltaic energy is inevitably and intimately intertwined with the uses we choose to make of it. The future depends on our ability to reach imaginative solutions as the generation of power is decentralized and utilities ease away from their role as suppliers to assume more the role of brokers.

The stakes are very high, and we cannot wait until every aspect of the technology is completely ready for commercialization. Photovoltaic power and the changes it will bring are inevitable. We would be well advised to deal with these realities affirmatively and deal with them now.

As the reins of government pass to a new administration that is strongly committed to increasing the national energy supply, we are face to face with a watershed decision on a new energy technology. Photovoltaic technology is no longer exotic, casting a distant rosy glow over the 21st century. We must begin to treat it as a practical here-and-now reality to be integrated fully into our decisions about energy supplies.

FURTHER READING

Application of Solar Energy to Today's Energy Needs. Vols. I, II. U.S. Congress, Office of Technology Assessment. Washington, D.C.: U.S. Government Printing Office, 1975.

Commoner, Barry. "The Solar Transition." *New Yorker,* Part I: April 23, 1979; Part II: April 30, 1979.

Environmental Quality: The Ninth Annual Report of the Council on Environmental Quality. U.S. Executive Office of the President, Council on Environmental Quality. Washington, D.C.: Government Printing Office, 1978.

---

[15] Dual-purpose systems collect heat energy in addition to providing electricity. For this reason, they are sometimes called "total energy" systems.

Lovins, Amory B. "Energy Strategy: The Road Not Taken?" *Foreign Affairs* 55, October 1976, pp. 65–96.

Schumacher, E. F. *Small is Beautiful. Economics as if People Mattered.* New York: Harper & Row, 1973.

Stobaugh, Robert, and Yergin, Daniel, eds. *Energy Future: Report of the Energy Project at the Harvard Business School.* New York: Random House, 1979.

Wilson, Carroll L., project director. *Energy: Global Prospects, 1985-2000—Report of the Workshop on Alternative Energy Strategies, Massachusetts Institute of Technology.* New York: McGraw-Hill, 1977.

CHAPTER 2
Technology

The photovoltaic phenomenon—the process by which light is converted silently and directly into electricity, without the elaborate machinery we usually associate with the generation of electricity—is elegant. It is certainly far less complicated than an atomic reactor, or, for that matter, even a conventional coal- or oil-fired plant. A review of the science underlying photovoltaics will help to explain why photovoltaics is at once safe, clean, durable, reliable, energy-efficient, and increasingly socially and economically attractive. A background in high school or college physics is quite adequate to appreciate the nature of the processes involved.

LIGHT

Light drives the photovoltaic process and provides the energy that is translated into electric energy.

Light is composed of tiny bundles of energy which act like individual bullets. These bundles have mass (weight), and they travel at extremely high, yet finite, measurable speed. Each bullet is called a photon, and the energy a photon possesses is the product of its mass and speed, just as is the case for a bullet from a gun. A stream of photons—a beam of light—behaves in some ways like a string of waves, and for decades light was thought to be a special kind of wave motion in a mysterious hypothetical medium called the ether. The apparent length of these waves is shorter for photons of higher energy, and, in fact, wavelength is one way of indicating energy content. More than a hundred years ago, a French scientist (Becquerel, in 1839) noticed that when light fell on one side of a very simple kind of battery cell, it produced an electric current in a wire. For many years no one knew why. Years later the mystery began to unravel when it was discovered that an atom consists of a minute nucleus surrounded by electrons. When a photon strikes an atom, it can interact with

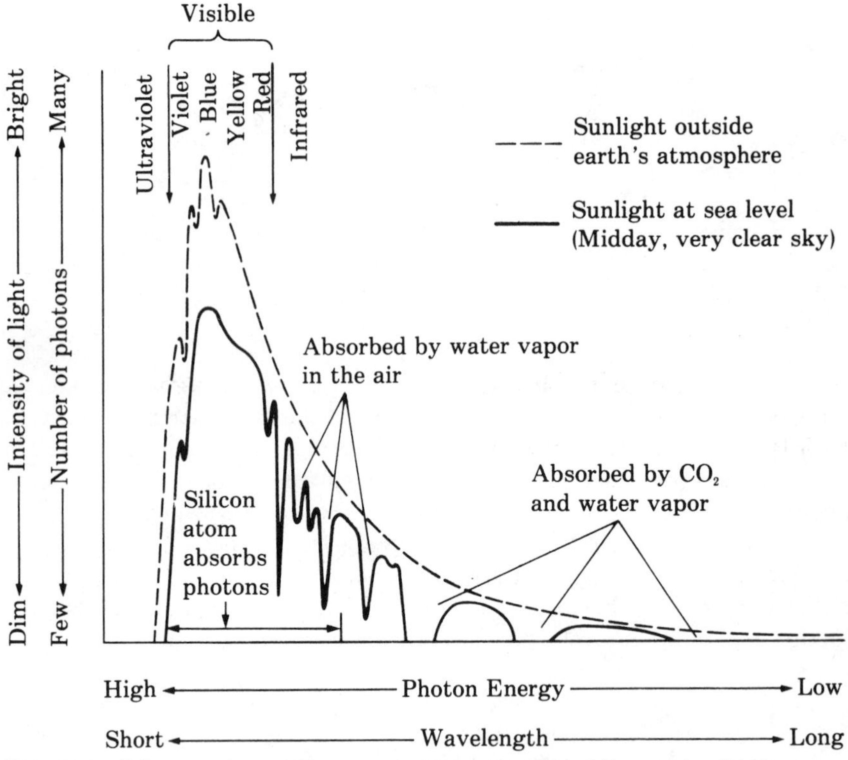

Fig. 2–1 Solar spectrum. The area between the dotted line and solid line represents the the portion of sunlight absorbed by the earth's atmosphere; some components of sunlight absorb strongly in certain regions. Our eyes are sensitive to the part of the spectrum where the photons are most numerous. A silicon atom will absorb photons above the energy level shown.

the electrons and be absorbed. The added energy can drive one of the atom's outer electrons off. (An electron freed in this manner is known as a conduction electron because it is free to move and form an electric current.)

This interaction between light and electrons is at the heart of all photovoltaic devices. The energy being continually generated within the sun, whose surface is about 10,000 degrees Fahrenheit (6,000 degrees Celsius) is liberated as a stream of photons of various energies leaving the surface. Indeed, because of this, the sun is slowly losing weight, or mass. Each photon reaching the earth has all of the energy it possessed when it left the sun's surface some 8 1/2 minutes earlier. A simple picture-diagram of sunlight shows the distribution, or spectrum, of these energies and how they relate to the colors our eyes detect (Fig. 2–1).

In solar photovoltaics, two aspects of sunlight are important: how many photons reach a given spot on the earth, and how much energy they have. As we suggested, sunshine is like a stream of bullets, all traveling at the same speed, some large and some small. Since energy is the product of speed times mass, the photons are distributed over a range of energies.

Above the earth's atmosphere, the energy the earth receives is about 1,358 Watts per square meter.[1] At sea level, this energy is reduced by atmospheric attenuation to about 930 Watts per square meter. (A square meter contains 10.57 square feet, an area somewhat larger than one square yard.) Thus, on a clear day when the sun is directly overhead or nearly so, if all of the sun's energy falling on one square meter of the earth were converted into electricity, it would light nine 100-Watt bulbs. Unfortunately it is not possible to convert even half of this solar energy into electricity. Nevertheless, the total energy being received is large enough that practical conversion devices are now becoming economically attractive.

About 30 percent of the energy reaching the earth's surface is immediately reflected into space as earthshine.[2] Another 47 percent is absorbed, at which instant the photons' energy is changed into heat.[3] (This is why sunlight feels warm on the back of your hand even on a cold day.) Indeed, photons must be absorbed for heat to be released. In terms of heat energy, one square meter of full sunlight at high noon yields about 750 calories, or about 3,000 Btu's every hour.

SINGLE CRYSTAL SILICON

The material most commonly used to make photovoltaic converters is silicon. Silicon is a good photovoltaic material because most of the incoming photons will displace electrons. All of the photons with energy greater than 1.08 electron Volts (corresponding to a wavelength of 11,500 Angstroms, in the infrared) will be absorbed and activate an electron. In the diagram of the solar spectrum, it appears that nearly all of the energy in sunlight could be converted to electricity. Unfortunately, this turns out not to be the case, because one photon can trigger the release of only one electron. Any photon that has more energy than the minimum required for

---

[1] Terms not explained in the text are defined in the Glossary.

[2] A. L. Hammond, W. D. Metz, and T. H. Maugh II, *Energy and the Future* (Washington, D.C.: American Association for Advancement of Science, 1973), p. 49. See also Appendix 7.

[3] Of the remaining energy, close to 23 percent goes to evaporation of water, precipitation, etc. A relatively small amount (less than 1 percent) goes to wind, waves, and photosynthesis. See Appendix 7, "Earth's Energy Balance."

releasing one electron has its remaining energy converted to heat in the silicon.

The net effect is that silicon solar cells can at best convert about 23 percent of the total energy in sunlight into electricity. This is a theoretical limit imposed by nature's building blocks and our ability to build practical conversion devices. In some cases it may be profitable to build systems which capture this excess heat energy for practical applications where both electricity and heat are needed. Such combined systems are called photovoltaic-thermal, or "total energy" systems.

There are other inefficiencies in practical silicon-based systems that further limit the amount of electricity that can be produced. First, some sunlight never gets into the cell because it is reflected from the surface. This loss can be partially overcome by using antireflective coatings. Second, the wire net or grid used in the construction of the cell blocks out some incoming sunlight. Third, depending on the purity of the silicon and the refinement of the crystal structure, there are internal losses of energy. And, finally, in a photovoltaic array all of the area devoted to the collection of sunlight is not fully covered by solar cells. While this is minimized by making the cells rectangular or hexagonal rather than circular, as most early cells were, some space is still not used. For silicon systems, these inefficiencies typically may be:

Energy Loss Due to	*Percent of Sunlight Not Converted into Electricity*
Overly energetic photons	32
Under-energetic photons	24
Other internal cell losses	21
Front surface reflection	3
Shading by contact bar and fingers	3
Cell packing density	2
Total	85

Good silicon photovoltaic systems for sale in 1980 had an overall conversion efficiency of about 15 percent.

Let's look at the basic properties of matter that the photovoltaic phenomenon capitalizes on. The simplest atom is hydrogen, which has one proton for a nucleus and one electron in orbit around it. The largest atom found in nature in any appreciable quantity is uranium, which has 92 electrons in orbit around a nucleus of protons and neutrons. Beginning with hydrogen, and moving toward larger atoms of heavier elements, the number of electrons builds up around the nucleus in successive layers or shells. The outer shell in the smaller atoms is full if it contains eight electrons; the addition of another electron starts a new shell (Fig. 2–2).

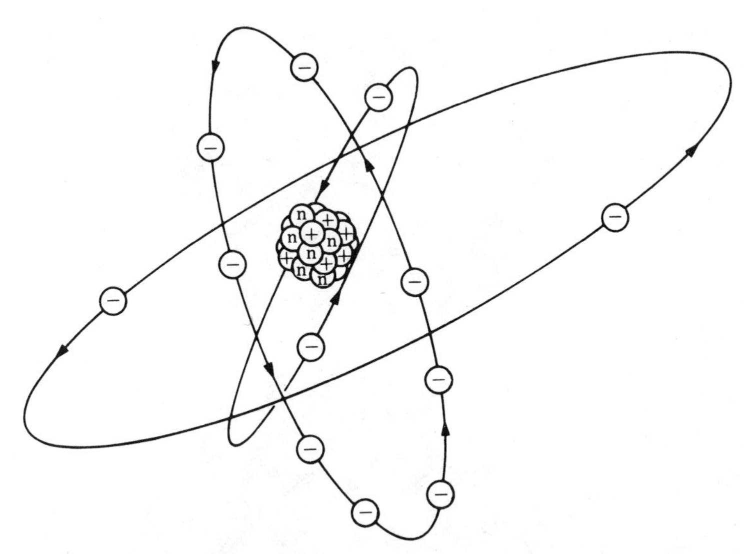

Fig. 2–2 Magnesium atom. One of the smaller atoms is magnesium, which is metallic. The nucleus is a tiny cluster of 12 protons $\oplus$ and 12 neutrons ⓝ. The elliptical orbits around the nucleus (in three dimensions) are the successive layers, or shells, of electrons $\ominus$. The innermost shell of all atoms is unique in that it is filled by only two electrons.

Each proton has one positive charge, each electron one negative charge. Every atom has the same number of electrons as there are protons, so that the atom as a whole is electrically neutral. Neutrons carry no charge.

In metals, such as magnesium, the electrons in the outer shell are not strongly attached, moving easily from one atom to another. For this reason, metals are good conductors of electricity, which is merely a flow of these outer shell electrons.

The silicon atom has 14 protons and 14 neutrons in the nucleus and 4 electrons in the outmost shell.

An electron is a minute bit of matter compared to the atoms of which it forms a part. It has a mass about 1/1,840 that of a single proton or a single neutron, both of which compose the nucleus of an atom. If the outer shell of an atom contains only one or two electrons, these are loosely held and move easily from one atom to another. Elements composed of such atoms are metallic and readily conduct electricity—which is itself nothing more than a flow of these outer shell electrons. (Indeed, a wire conductor can be likened to a pipe full of water: if additional electrons are pushed in one end, the electrons already in the wire tend to be forced out the other end, or at any point that they can find an escape route. A generator is merely a device

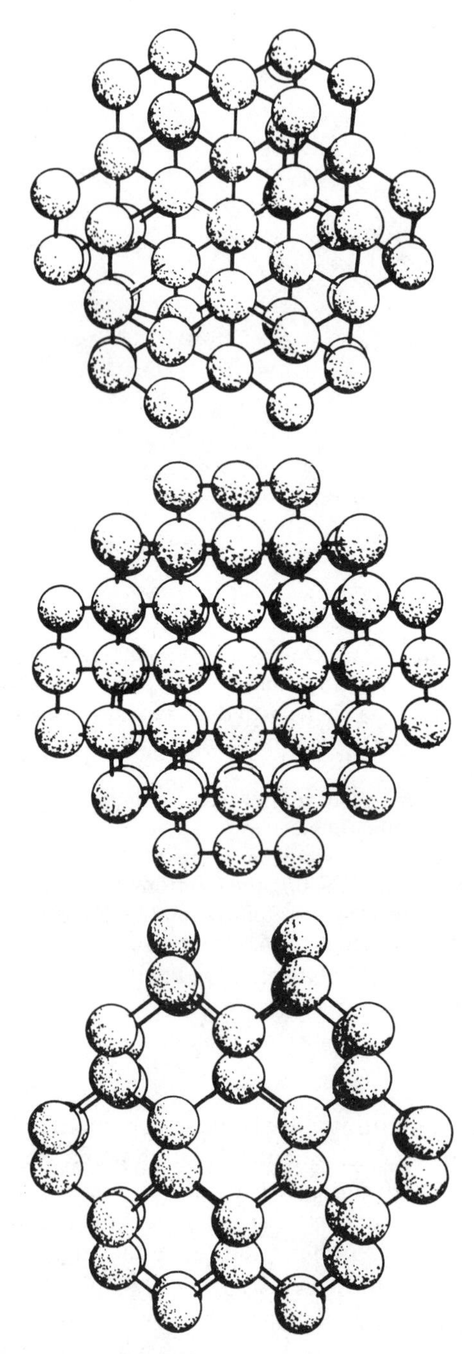

Fig. 2–3 Model of a silicon crystal. Each atom is firmly positioned by electron bonds to four surrounding atoms. This model is shown from several angles.

that provides the push to the electrons already in the wire; what it generates is flow, or current, but it does not create electrons or their electrical charges.)

On the other hand, if the outer shell of an atom contains six or seven electrons, the material is non-metallic and a poor conductor of electricity because the electrons are tightly bound.

Between these extremes are elements whose atoms have three, four, or five electrons in the outer shell. These outer electrons can be freed, but only if given some additional energy, or push, from outside. Because they conduct electricity slightly, such elements are called semiconductors. An atom of silicon, a principal constituent of sand, has four outer shell, or valence, electrons.

A silicon atom readily absorbs a photon, and the added energy will "excite" or activate one of the outer electrons and thus free it. This is best accomplished when silicon atoms are lined up in precise rows or positions. This is called the crystalline state. In any crystal, the atoms or molecules are arranged in perfect geometrical formation (Fig. 2–3). (The opposite is the amorphous state, in which atoms or molecules are jumbled together in no regular pattern.)

When a photon strikes this silicon crystal, it penetrates until absorbed by an atom (Fig. 2–4). Almost immediately the photon's energy is transferred to one of the outer electrons, which then breaks free from the atom and bounces away through the crystal lattice leaving a vacant position, or hole, where it was. When a beam of light falls on a silicon crystal, millions of electrons are freed in this manner, and an equal number of holes are left in the lattice. Any electron that happens upon any hole may refill it, giving off a minute amount of heat in the process. When the light is shut off, all the displaced electrons are quickly reabsorbed in the holes. The crystal returns to its original state, becoming slightly warmer. Because the atoms do not change position, the crystal itself remains completely unaltered in size, weight, form, and appearance.

A trick has been discovered that upsets this orderly state of affairs for a useful purpose. By introducing a few foreign atoms into a silicon crystal, the perfect arrangement of outer shell electrons can be upset or disordered. A crystal can be created which has either too many electrons or not quite enough to satisfy its internal structural balance. Two elements are typically used for this purpose: boron, which has three outer shell electrons, or phosphorus, which has five. Only a minute amount of these impurities is needed, perhaps one part in a million.

Whenever a boron atom is added, it leaves a vacancy of one electron in the lattice. Thus, a silicon crystal "doped" with boron needs electrons to make the lattice complete, and it tends to absorb electrons much as a

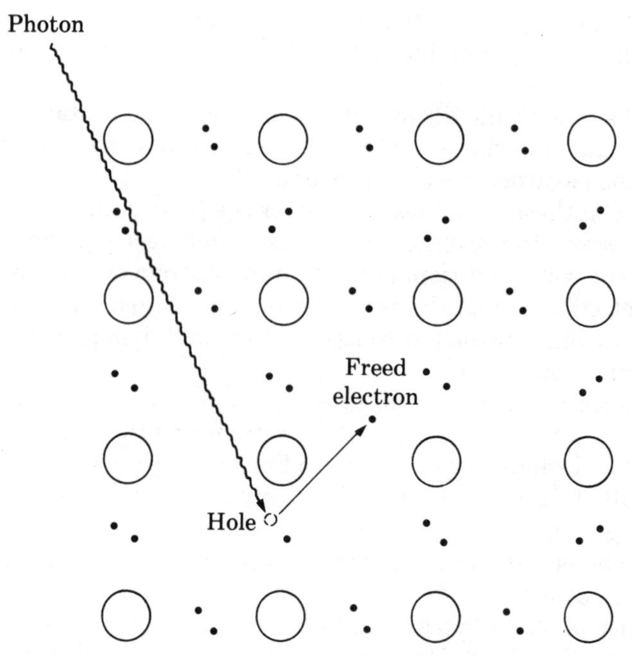

Fig. 2–4 Silicon crystal. When molten silicon cools very slowly, the atoms (large circles) align themselves in a perfect cubic pattern. Each atom has four outer shell electrons (black dots), which are shared with a neighboring atom to provide a stable structure (shown here in two dimensions). A photon (particle of light) is shown freeing an electron.

sponge soaks up water. On the other hand, silicon doped with phosphorus atoms, each of which has five outer shell electrons, will have an excess of electrons in the lattice.

HOW A PHOTOVOLTAIC CELL WORKS

A solar cell is made by placing a thin layer of phosphorous-doped silicon in intimate contact with a layer of boron-doped silicon. When light falls on the cell, photons are absorbed and electrons are set free. The excess electrons accumulate in the phosphorous-doped silicon, which is called n-silicon because electrons have negative charge. If one end of a wire is attached to this top layer and the other end connected to the layer beneath, electrons will leave the upper layer, flow through the wire, and be absorbed by the boron-doped silicon, which is called p-silicon, meaning positive (Fig. 2–5).

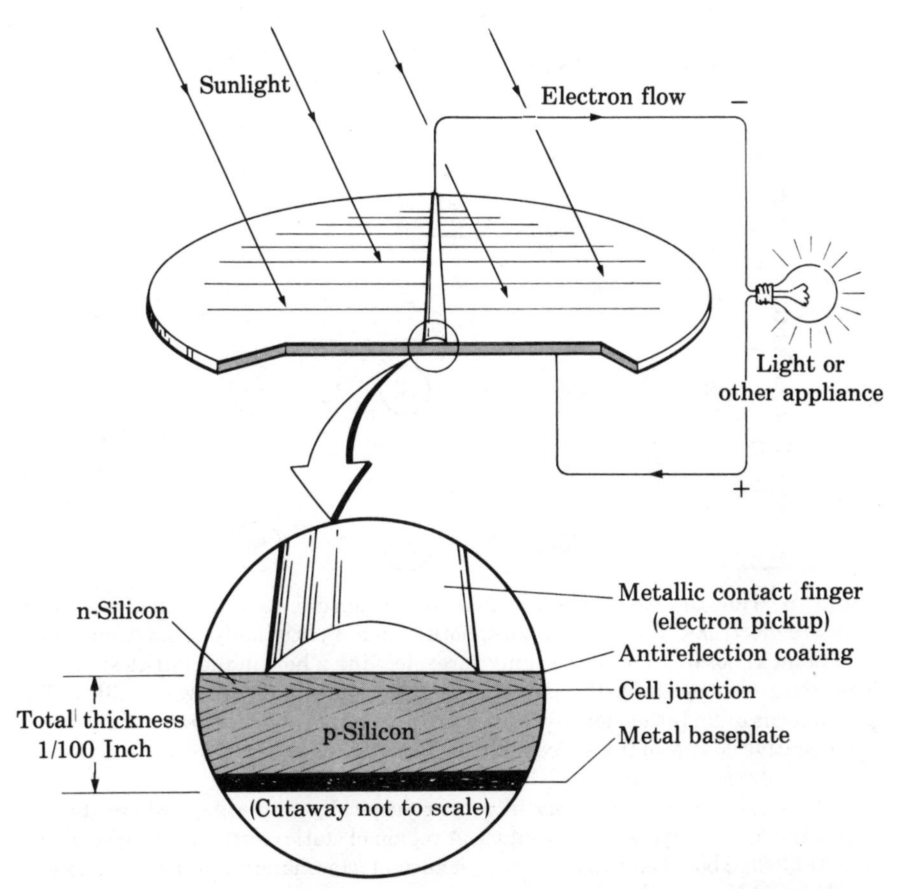

Fig. 2–5 Schematic diagram of single crystal silicon solar cell. Sunlight penetrating into the silicon frees electrons, which filter upward through the cell junction, concentrating in the n-silicon layer. From there the electrons flow through the metallic fingers into the wire, finally completing the circuit by returning through the baseplate to the p-silicon layer.

The thin crystal silicon wafers, sawn from a solid ingot, are typically 3–6 inches in diameter. This was one of the earliest commercial solar cells, widely used on space satellites since the 1950s. Silicon cells of this general type still constitute the mainstay of the commercial solar photovoltaic business, although now rectangular in shape in order to provide more compact arrays.

The electrons flowing through the external wire circuit can be used to light a light or turn a motor, just as electricity from any other source. Thus, the solar photovoltaic cell, using light only, generates electricity.

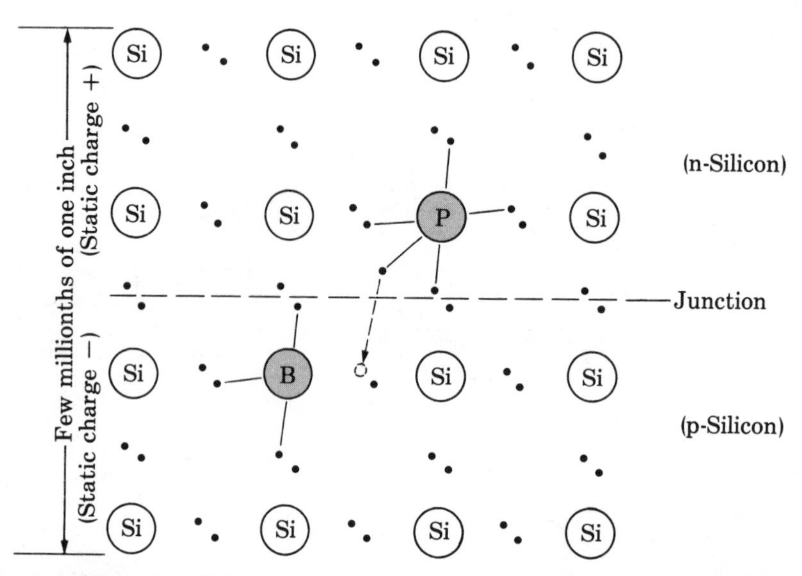

Fig. 2–6 How the cell barrier is established. Open circles are silicon atoms; black dots are electrons. The neutral phosphorus atom, P, originally has five electrons. The neutral boron atom, B, has only three, leaving a hole in the lattice structure. The extra electron from the phosphorus atom immediately moves to fill it. The phosphorus atom is then left with a net positive charge and the boron atom picks up the negative charge of the added electron. No light (photon) is involved in this step, which occurs spontaneously.

This exchange occurs only in a microscopically thin region where the two oppositely doped layers are in contact. A region of static electrical charge can also be established between two dissimilar semiconductor materials or between a semiconductor and a metal.

Why do the free electrons prefer to leave via the wire rather than be absorbed in the holes they have left behind in the crystal? When the cell is made, a peculiar phenomenon occurs at the plane of contact between the two silicon layers. Immediately, some of the excess electrons in the n-layer diffuse a short distance across the interface into the p-layer just beneath. They are attracted by the holes there, which they "want" to fill. This leaves the phosphorus atoms in the n-layer without enough electrons to balance fully the positive charge in their nuclei, while too many electrons are located in the p-layer around the boron atoms. A very thin layer of static electrical charge is thus formed along the zone of contact, or junction, between the two layers of silicon (Fig. 2–6).

Such static electricity is a commonplace phenomenon. If you rub your hand hard across your hair, some electrons from the hair atoms are rubbed off and stick to the palm of your hand. If you then hold your hand close to your hair, hair will rise toward it because the hair, now positively charged, is attracted by the negative charge on your hand. A sweater taken off suddenly will often pick up enough electric charge so that, in a dark room, you can see sparks jump from it when it is laid down. Clothes tumbling in a dryer develop strong static electricity as electrons are rubbed off one fabric by another. Such effects are more noticeable in dry weather when the air will not conduct the electrons back to their point of origin as readily.

Because electrons are charged particles, it is difficult for them to pass through the zone of static electric charge in a solar cell. For this reason, the charged region is called the cell barrier. When a cell is fabricated, the barrier establishes itself instantly. It lasts for the life of the cell. It never wears out. The cell barrier is extraordinarily thin, calculated to be about one-millionth of an inch thick. Every solar cell must have such a barrier, or zone of static charge. No solar cell would work without it.[4]

Because the barrier resists the passage of electrons through it, only electrons with high energy (great speed) can penetrate it. As a result the barrier acts as a filter that lets high energy electrons through and stops low energy electrons.

An analogy can help explain how this happens. A solar cell can be likened to two tables with marbles on them (Fig. 2–7). The two tables are connected by a small elevation or ridge over which marbles can move if they have enough speed. Table A on the left is vibrated gently (in the horizontal plane) so that the marbles are jostled about, and occasionally a marble will roll over the ridge to Table B. On the other hand, Table B is vibrated forcefully so that frequently marbles will have enough energy to get over the ridge in the other direction to Table A. Eventually, as marbles accumulate on Table A and the number remaining on Table B diminishes, the number of marbles crossing the ridge will be the same in each direction, and a condition of equilibrium will be established. However, if the two tables are connected by an external channel, which we will assume is flexible, the press of marbles on A will cause some of them to be pushed through this channel back to Table B.

In a solar cell, the p–n junction is like the ridge between the tables. The free electrons act like the marbles. The p-silicon lattice, corresponding to Table B, has an excess of holes due to the boron dopant and these holes tend to absorb electrons. As a result, there are not many free electrons in the p-silicon, but those that are free move at high speeds (have high energy).

---

[4] This zone of static electricity is sometimes called the "space charge." It is also referred to as the "depletion zone" because it is devoid of both free electrons in the n-layer and of holes in the p-layer since the electrons have been absorbed by the holes.

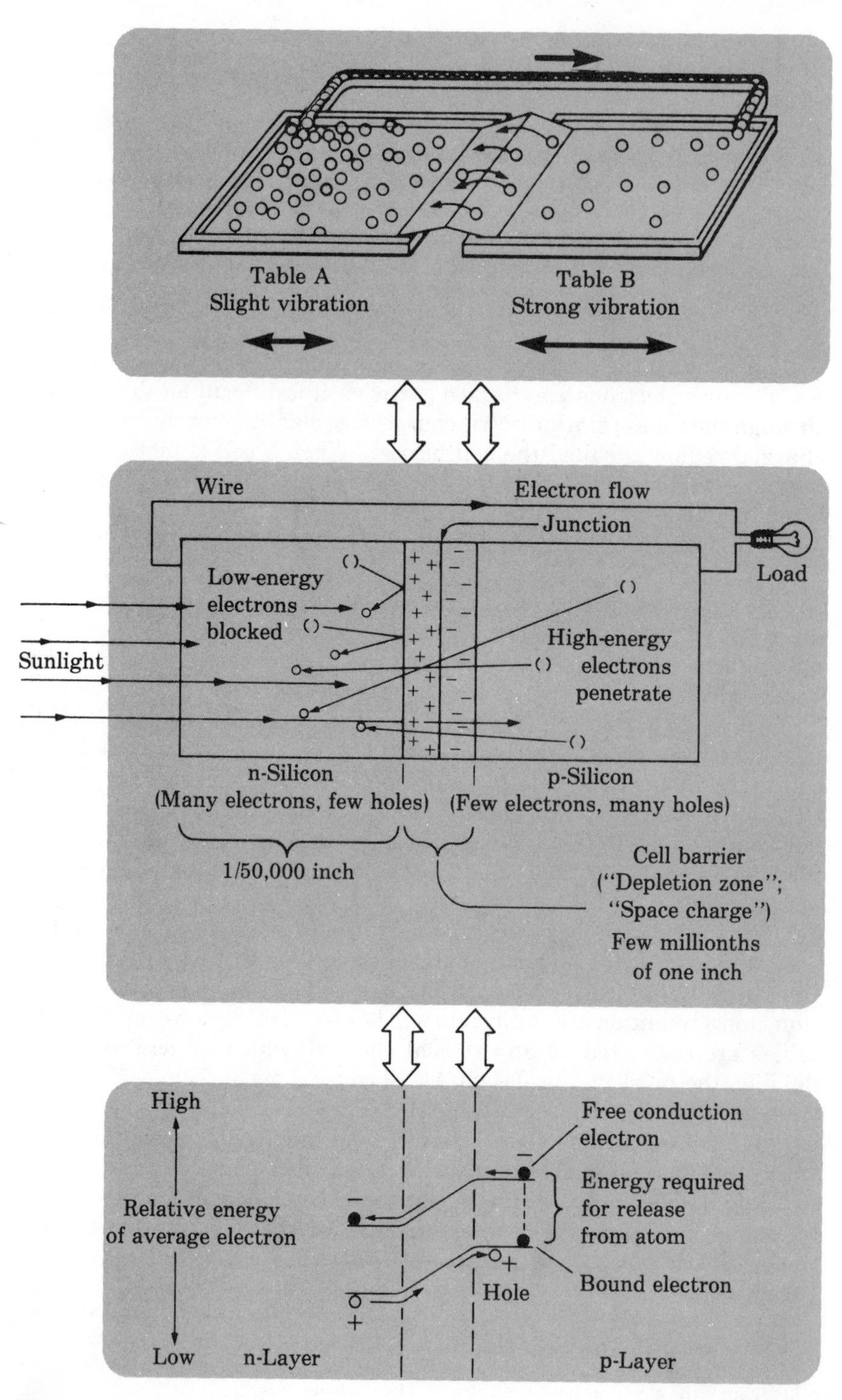
Table A
Slight vibration
Table B
Strong vibration
Wire
Electron flow
Junction
Load
Low-energy
electrons
blocked
Sunlight
High-energy
electrons
penetrate
n-Silicon
(Many electrons, few holes)
p-Silicon
(Few electrons, many holes)
1/50,000 inch
Cell barrier
("Depletion zone";
"Space charge")
Few millionths
of one inch
High
Relative energy
of average electron
Low
n-Layer
p-Layer
Free conduction
electron
Energy required
for release
from atom
Bound electron
Hole

The opposite is true in the n-silicon, which corresponds to Table A. Here there are not enough holes and there are too many electrons, so that the average electron has less energy (is moving slower) than those in the p-silicon. Electrons in both layers are moving at random, some driving by chance into the barrier. The higher energy ones from the p-silicon penetrate the barrier into the n-silicon, while low energy electrons in the n-layer are prevented from returning. Thus voltage, or electron pressure, is created between the two layers. If a wire is connected externally from the n-layer to the p-layer, the excess electrons will readily flow through it.

As electrons migrate preferentially into the n-layer, the holes in the lattice, of course, can be visualized as progressively "moving" in the opposite direction. Electrons coming back through the external wire eventually end up being absorbed in holes at the base of the p-layer, which in effect were originally produced when the electrons broke free. Electrons move so rapidly that if the source of light is shut off, the free electrons are almost instantly reabsorbed in holes and the process comes very quickly to a halt.

The incoming photons keep the process going by continually creating new electron-hole pairs. Indeed, it does not matter which side of the junction the new pairs are created on. It is important only that they be created in the vicinity of the junction. If a photon is absorbed in the n-layer, the electron that is freed simply adds to the number already present on that side, which is desirable; the hole migrates eventually through the junction as incoming electrons successively fill it. If a photon is absorbed in the p-layer, the process is as described above.

Fig. 2–7 How a solar cell works. In this diagram, the solar cell is pictured (center) lying on its side with the sunlight entering from the left. At the center is the cell barrier—also known as the depletion zone, or space charge. This is a thin region of static electric charge which resists the passage of electrons. High-energy electrons on the right are able to penetrate this region; low-energy electrons on the left are repelled. The process can be pictured as basically mechanistic, as depicted by the table of marbles (top). Just as the marbles accumulate preferentially on Table A, so the electrons accumulate in the n-silicon.

In the cell (center), the free electrons bounce about at random, colliding frequently and changing directions with every collision. Thus the long, straight paths, shown here to illustrate the principle, are an oversimplification.

The physicist uses an energy diagram (bottom) to portray the movement of electrons from a high-energy state (right) through the barrier to a lower-energy state (left), while an equal number of lattice vacancies migrate in the opposite direction.

The n-layer is usually made of such thickness that the junction will be at optimum depth for absorbing photons. For silicon, this is about .5 micron (1/50,000th of an inch).

The entire thickness of the cell need be no greater than 250 microns (0.25 millimeter), about 1/100th of an inch. This is only three times the thickness of a human hair. It is theoretically possible to make solar cells that are extraordinarily thin, and thus much cheaper, especially in mass production.

To portray what is happening in a solar cell, the physicist resorts to a diagram as shown in Figure 2–7. Here the cell is imagined to be in the position of the two tables of marbles, with the n-silicon on the left. Like the ridge between the tables, the junction or barrier is in the middle. The physicist uses the vertical dimension in the diagram to portray the relative energy of both the free electrons and the bound electrons in each region of the cell. The free or "conduction" electrons, represented by the upper line, move from the region of higher average energy in the p-silicon on the right, through the barrier, to the region of lower average energy in the n-silicon on the left. Simultaneously, the holes or vacancies migrate in the opposite direction.

From the standpoint of electrical theory, the electrons in the p-layer near the junction are at higher voltage than those in the n-layer just across the junction. Because, by definition, electricity flows from higher voltage to a lower voltage, a current flows through the junction region from the p-layer to the n-layer. Indeed, the explanation is frequently given that the electrons in the p-layer near the junction are "sucked down" through the barrier to the zone of lower voltage.[5]

Particle physics sees the matter a bit differently. The process can be envisioned as basically mechanistic. The photovoltaic phenomenon is remarkable in that here electrons—the subatomic particles that constitute an electric current when they are flowing in unison in one direction—are behaving more like the bits of matter that they are. Within the crystal, they are diffusing through the cell barrier, much as molecules of a gas might diffuse through a porous barrier. But because these are electrons in motion, their motion does in fact conform with electrical theory, and the process is often explained in that context. Thus the holes, which are regions of positive electrical charge, are pictured as "moving" in the opposite direction from the negatively charged electrons, in order to make things come out even from the standpoint of electrical theory.

---

[5] Another explanation sometimes given is that the cell barrier separates new electron-hole pairs as they are created by photon activity. This is the net result, but the picture is rather strained.

It is similar to saying that when automobiles move from the city to the suburbs in the evening, an equal number of empty parking spaces "flows" into the city. If each driver had several errands to perform and therefore made a number of stops on the way home, occupying a succession of parking spaces, then the picture of empty spaces flowing toward the city as cars flowed homeward takes on more realism.

The electronics engineer uses the flow of electrons and of holes to describe what is happening in a photovoltaic cell because the lattice vacancies are regions of positive charge which, in his system of accounting, must exactly equal the number of negative charges carried by the moving electrons. But in terms of movement of particles of matter, only the electrons are involved.

The sole function of the photons is to keep replenishing the supply of free electrons by energizing electrons from their trapped positions in orbit about the nuclei into a free state, called the conduction band. The cell does the rest of the job. The photons, in effect, provide a pumping action, and the energy they supply is converted into the energy of electrons in motion, and thus into an electric current that lights a light, drives a motor, or produces heat.

The amount of current (amperage) produced by a photovoltaic cell is proportional to the amount of light falling on the cell (the number of photons entering it). For this reason, current increases with the area of the cell as well as with the intensity of the light. The voltage, on the other hand, depends on the materials used. All silicon cells produce about 1/2 volt regardless of cell area. A single-cell flashlight battery is rated at 1½ volts, which can be provided by three silicon photovoltaic cells connected in series so that their voltages are additive.

Solar cells are unique energy producers in that no materials are consumed or given off. For this reason solar photovoltaic receivers can be completely sealed units that, in principle, are as durable as the materials from which they are made. In this important respect, photovoltaic devices are very different from batteries. The cell material in batteries must undergo change in chemical composition, during which the atoms, in the form of charged ions, move about. For example, lead may dissolve in acid as electricity is withdrawn and be plated out again as the cell is recharged. Because the material itself changes form, it never is replaced exactly as it was, and even the best batteries in time simply wear out.

But in photovoltaic cells, the atoms themselves do not change position; only the electrons between them move. The material does not change shape or form any more than, say, a copper wire that conducts electricity for years without undergoing any change. It is possible to make solar cells that operate for twenty or thirty years with very little loss of effectiveness.

Energy from light is simply being converted directly into electrical energy for external use.

POLYCRYSTALLINE SILICON

Other forms of silicon that are much cheaper to produce than large single crystals will also work in photovoltaic applications. When molten silicon cools under normal circumstances, the resulting ingot consists of a myriad of microscopic crystals. If the cooling is done very slowly, the crystallites will be larger in size. The product, called polycrystalline silicon (polysilicon or poly) is granular in nature, and the size of the crystallites depends on the conditions under which the silicon cooled. Polysilicon can be formed as an ingot, or solid block, or in the shape of a ribbon drawn up from a molten bath, or by boiling the silicon off and allowing the vapor to deposit and cool on some flat surface. These methods are much less expensive than growing single crystal silicon.

Silicon cells made from polycrystalline silicon are somewhat less efficient in converting sunlight into electricity. Minute electrical shorts occur along the grain boundaries between the crystallites. The larger the crystallites, therefore, the fewer the boundaries, and the more the resulting cell behaves as if it were made from a single crystal (Fig. 2–8).

In general, it is desirable to have the individual crystals much longer than the thickness of the cell in order to reduce losses from electrical shorts along the grain boundary. Grains ranging in size from about 1/10 millimeter to several millimeters in length can be made, and they work quite satisfactorily.

One way of reducing electrical seepage along the boundary is to treat the tiny crystals with some material that, in effect, insulates the edges of the grains. Hydrogen is being used for this purpose. Cell efficiencies as high as 10 percent are being achieved for polysilicon, depending on the design of the cell. Polycrystalline silicon grown in ingot form still has to be sawed into thin wafers, an expensive step required in making cells from single crystal material. But polysilicon can now be formed in a thin sheet in the first place, by vapor deposition or by simply dipping a substrate, such as a piece of graphite, in a bath of molten silicon, eliminating the sawing step. Such procedures, which are being developed vigorously, are more adaptable to mass production techniques and should make it easier to produce cells that are larger in area. The goal is to reduce cell cost without sacrificing conversion efficiency, reliability, or the useful lifetime of the product.

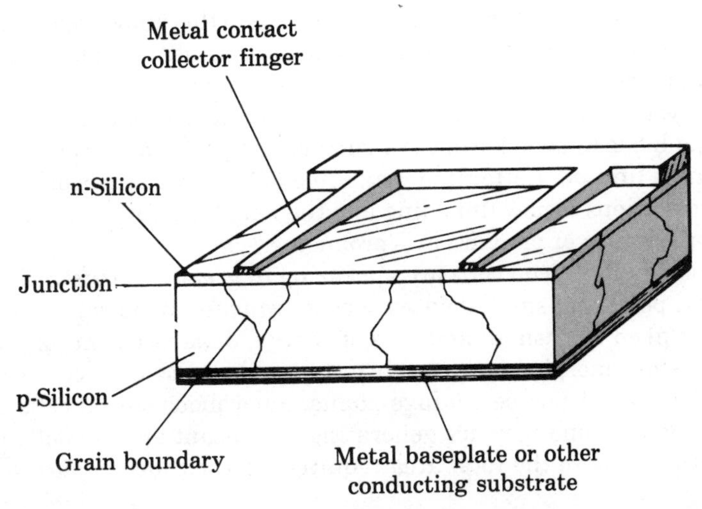

Fig. 2–8 Cross-sectional diagram of silicon cell made of polycrystalline material. Free electrons find conducting pathways along the sides of the crystallites, enabling them to recombine with holes before passing through the cell junction. In addition, electrons in the n-layer are inhibited from moving across crystallite boundaries to reach the metallic grid fingers on top. Thus the cell is partially shorted out internally. This problem can be reduced in two ways: by providing larger crystallites, which reduces the number of grain boundaries; and by treating the silicon with hydrogen gas, which is absorbed within the cracks and, for reasons not yet fully understood, reduces the unwanted conductivity of electrons.

AMORPHOUS SILICON

Even amorphous silicon, the noncrystalline material (also called alpha-silicon) can be used to make solar cells. Amorphous (without form) silicon is pure silicon that has no crystal properties. The silicon atoms in the solid are randomly distributed. Glass, for example, is amorphous, not crystalline, material. Very thin films of amorphous silicon can convert sunlight into electricity. Since a very small amount of material is used and the crystallization step is avoided, amorphous silicon cells are very inexpensive.

The light energy conversion efficiency of amorphous silicon, however, is much lower: typically, 3–6 percent, compared with 15–16 percent for single crystal cells. When solid state theory was developed, it centered on the structure and behavior of individual atoms and the forces that bind individual atoms together. It was easy to extend this knowledge to crystals because crystals are perfect geometrical arrangements of atoms. Because there was long range order, what applied to the smallest unit applied to the whole crystal. The behavior of amorphous materials is much more complex, and basic knowledge of the structure and performance of amorphous material is limited. Better understanding of how matter performs in such conglomerations is now unfolding and the long term future of amorphous materials for power production is promising.[6]

Amorphous silicon cells are appearing in great numbers in Japanese products, powering small devices such as watches, hand calculators, and toys that need very small amounts of energy. Japanese firms are leading producers of amorphous silicon. If techniques can be devised for raising their efficiency a few percentage points, amorphous silicon cells may in time become competitive for generating significant amounts of power in applications where the total area required for the collector surface is not critical.

OTHER KINDS OF SOLAR CELLS

Solar cells can be made from a number of materials other than silicon, and formed in a variety of designs. Cells are classified both by material and by the type of junction, or barrier, they employ.

There are four basic kinds of junctions. In the silicon cell, the junction is the very thin region at the boundary between the phosphorous-doped and the boron-doped portions. The basic material is all silicon, and such a junction within a single material is called a *homojunction*. Solar cells can, however, be made of dissimilar materials. One type of solar cell has a layer of copper sulfide deposited on a layer of cadmium sulfide and the junction forms along the contact between the two substances. Where such dissimilar materials are used, the contact zone is called a *heterojunction*. A junction can also be established between a semiconductor material and a piece of metal, and this type is called a Schottky barrier, for its discoverer. Finally, in a metal-insulator-semiconductor, or MIS junction, a very thin layer (less than 0.003 micrometers) of some material such as titanium oxide is sandwiched between the metal and the semiconductor. A modification of the MIS cell is the semiconductor-insulator-semiconductor, or SIS cell. Each of these basic types of junctions offers various advantages and has associated limitations.

[6] For an excellent discussion, see David Adler, "Amorphous Semiconductor Devices," *Scientific American*, May 1977, pp. 36–48.

Some of the more promising semiconductor materials and their theoretical efficiencies for converting sunlight into electricity are:[7]

Material	*Maximum Conversion Efficiency, Percent*
Germanium	13
Cadmium sulfide	18
Silicon	25
Cadmium telluride	25
Indium phosphide	26
Gallium arsenide	27
Aluminum-antimonide	27

Acknowledging there are many reasons "that rule out any hope of reaching more than about four-fifths of these efficiencies," Bruce Chalmers of Harvard University pointedly adds, "Lest it be thought that an efficiency of less than 20 percent is unworthy of our technical sophistication, it should be remembered that the epitome of advanced industrial technology, the automobile, performs its energy conversion at an efficiency of less than 20 percent, and it performs its people-moving function at a far lower figure."[8] In generating electricity from fossil fuels, two-thirds of the energy is lost in the conversion process.

Extensive theoretical and applied research is in progress on these and a number of other materials. Depending upon the substances used, and the type of junction employed, literally dozens of different kinds of solar cells are possible. The choice of cell material and the design of the cell influence not only the cell efficiency, but also the fabrication method, which in turn affects costs. The worth of a cell, which is usually expressed in dollars per Watt of generating capacity, incorporates both factors.

Other than silicon, the materials most often used in solar cells today are cadmium sulfide and gallium arsenide. Cadmium sulfide cells are relatively easy to manufacture, inexpensive, and have achieved conversion efficiencies of about 8 percent. Depending on how it is made, the cell proper may be as thick as 35 micrometers or as thin as 5 micrometers. A metal substrate such as copper can be used, but more recently 1/8-inch-

[7] An excellent overview of the leading photovoltaic materials from the standpoint of cell efficiency is contained in "Photovoltaic Materials," an article by Evelio A. Perez-Albuerne and Yuan-Sheng Tyan, *Science* 208, 23 May 1980, pp. 902–907.

[8] Bruce Chalmers, "The Photovoltaic Generation of Electricity," *Scientific American,* October 1976, p. 39.

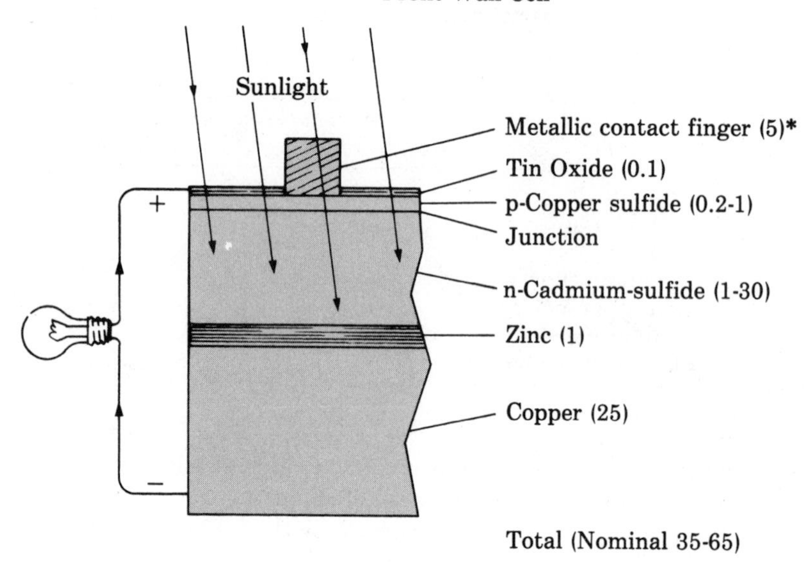
Front Wall Cell
Sunlight
+
−
Metallic contact finger (5)*
Tin Oxide (0.1)
p-Copper sulfide (0.2-1)
Junction
n-Cadmium-sulfide (1-30)
Zinc (1)
Copper (25)
Total (Nominal 35-65)

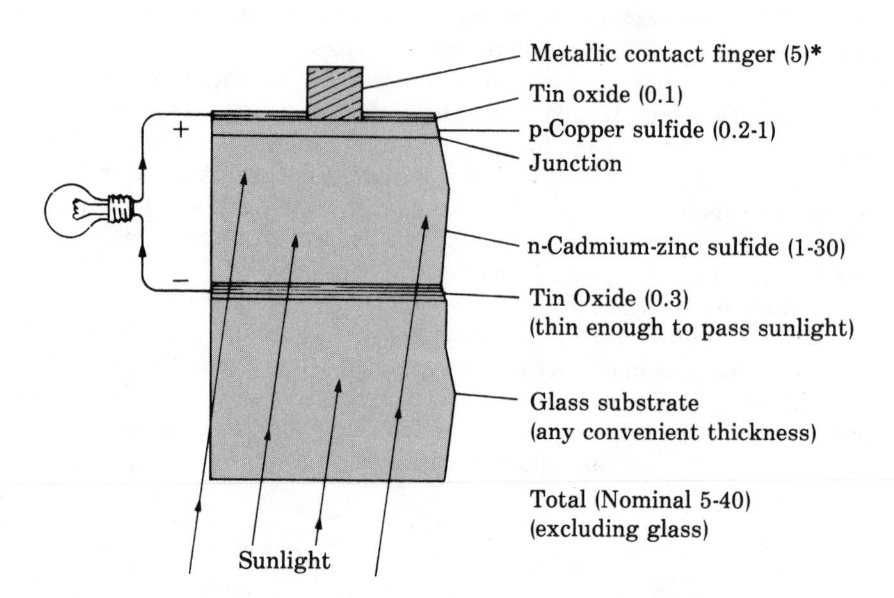
Back Wall Cell
+
−
Metallic contact finger (5)*
Tin oxide (0.1)
p-Copper sulfide (0.2-1)
Junction
n-Cadmium-zinc sulfide (1-30)
Tin Oxide (0.3)
(thin enough to pass sunlight)
Glass substrate
(any convenient thickness)
Total (Nominal 5-40)
(excluding glass)
Sunlight

thick (0.3 cm) glass panes are being used (Fig. 2–9). In production, the most expensive part of the completed product is the glass backing or substrate.

Gallium arsenide is important for two reasons. First, it is inherently more efficient because it captures a wider part of the solar spectrum than cells of other materials. Second, at high temperatures it does not lose efficiency as rapidly as silicon. Gallium arsenide is one of the best materials known for use in concentrator cells, which focus the relatively diffuse sunlight on a small area and produce high temperatures. Moreover, this material may be produced either in single crystal form or as a polycrystalline thin film. Efficiencies approaching 25 percent have been demonstrated with single crystal gallium arsenide cells, while 15 percent conversion has been attained with thin film gallium arsenide devices only 1–2 micrometers thick.

Closely akin to amorphous silicon are the so-called chalcogenide glasses. These are semiconductor substances that contain large amounts of sulfur, selenium, or tellurium. (A glass is a substance that has been cooled quickly so that orderly crystals are not formed. That is why glass breaks along curved lines rather than the straight, flat surfaces characteristic of crystals.) Solar cells can be made of these materials today, but their efficiencies are low. The future, however, may be another story.

As more is learned of the substances that make good solar cells and the arrangement of the atoms in such materials, chemists may be able to synthesize new compounds better tailored to specific requirements of photovoltaic systems. Vigorous research is underway on conductive polymers, plastics with interesting electrical properties.[9] This field is still in its infancy.

---

[9] J. Mort, "Conductive Polymers," *Science* 208, 23 May 1980, pp. 819–825.

Fig. 2–9 Distinctively different from silicon cells are cadmium sulfide-copper sulfide cells. Though less efficient than silicon cells, they are extremely thin and readily adaptable to mass production. Two types are shown. The Front Wall cell is made by depositing the required materials in successive layers on a thin sheet of copper. In the Back Wall type, materials are deposited on a pane of glass. Sunlight enters through the glass, which provides weather protection. The most expensive part of this cell is simply the one-eighth-inch-thick glass substrate.

Figures in parentheses are thicknesses in micrometers (1 micrometer is one-millionth of a meter, or about 1/25,000 of an inch).

In summary, although photovoltaic theory is complex, the phenomena involved are commonplace daily experiences—as familiar as the warmth of sunshine on the back of your hand, as the static electric charge in your hair, and the handy electric plug on the wall. There are no strange, new, unseen forces that society must try to comprehend.

The sheer elegance of the process is striking. There are no moving parts and no noise. In operation, nothing is consumed other than sunlight, and the output is high quality energy in the form of versatile electricity. Because there are no material by-products, there is no pollution.

Photovoltaics is distinguished from the other solar technologies by the opportunities it holds for further discoveries of a fundamental nature. By comparison, direct collection of solar heat and the indirect methods of tapping solar energy, such as wind, ocean thermal systems, and the use of biomass, are limited for the most part to engineering improvements and simplification of production procedures. There is less chance for significant scientific breakthroughs.

The photovoltaic process is now well understood; the research is probing a variety of collateral issues—the effects of grain sizes, impurities, and temperature; grain boundary effects; properties of dozens of new materials; and many cell designs.

FURTHER READING

Adler, David. "Amorphous Semiconductor Devices." *Scientific American,* May 1977, pp. 36–48.

Chalmers, Bruce. "The Photovoltaic Generation of Electricity." *Scientific American,* October 1976, pp. 34–43.

Fan, John C. C. "Solar Cells: Plugging into the Sun." *Technology Review* 80, August–September 1978, pp. 2–19.

Hovel, H. J. "Novel Materials and Devices for Sunlight Concentrating Systems." *IBM Journal of Research and Development* 22 (March 1978): 112–121.

Landsberg, P. T. "An Introduction to the Theory of Photovoltaic Cells." *Solid-State Electronics* 18 (1975): 1043–1052.

Neville, Richard C. *Solar Energy Conversion: The Solar Cell.* New York: Elsevier Scientific Publishing Co., 1978.

Pulfrey, David L. *Photovoltaic Power Generation.* New York: Van Nostrand Reinhold, 1978.

Sittig, Marshall. *Solar Cells For Photovoltaic Generation of Electricity—Materials, Devices and Applications.* Park Ridge, N.J.: Noyes Data Corp., 1979.

CHAPTER 3 Options

If price were no object photovoltaics would already be widely used as a clean, safe, reliable, silent source of elecric power. Price is indeed an object, however, as anyone who has studied his or her heating or electric bill and searched for alternative energy sources will agree. The price of photovoltaics is intertwined with the technology for producing photovoltaic cells. Genuinely economical photovoltaic systems depend on the successful development of new production methods that capitalize on economies of scale through automation, and the development of more energy-efficient collectors. Soon to be cost-competitive, photovoltaics can be expected to reorder the energy equation around which our economy is structured.

Photovoltaics becomes economic at different unit costs, depending on the application and relevant variables, such as climate, tax status of the owner, cost of money, and the price of utility-generated electricity. For a number of reasons, we use the figure of $.70 per installed Watt of peak capacity for the array as a benchmark at which photovoltaics becomes generally competitive for residential applications in the United States (all prices are in 1980 dollars).

PURIFYING SILICON

Most of the approaches to fabricating photovoltaic cells rely on silicon. Because a prime objective is to reduce costs, an adequate supply of low cost silicon of the required purity is a basic concern. The best source of silicon is silica, which is silicon dioxide, found in nature and throughout the United States as high grade sand. Quartz rock is another form of silica. To produce silicon, silica is fused to a melt and a controlled amount of carbon is added in a batch process. The carbon reacts to remove the oxygen as carbon dioxide, leaving relatively pure silicon, which cools into a block or ingot.

This medium-gray, metallic-looking substance is metallurgical grade silicon. Metallurgical grade silicon has about 1 percent impurities and sells for $.50–$1.00 per pound ($1.10–$2.20 per kilogram).

Converting this to semiconductor grade silicon is expensive, as impurities must be cut greatly. The silicon must be extremely pure (except for the minute, controlled quantities of dopants) because ordinary impurities upset the delicate electron structure that exists between the atoms. In fact, the required purity of semiconductor grade silicon is less than one extraneous atom per billion atoms of silicon. For comparison, in a world of some four billion persons this is less than one "impure" person per continent.

The required purity is reflected in a selling price for semiconductor grade silicon (in 1980) of $50–$70 per kilogram. At this price, the cost of silicon accounts for at least half the cost of the finished cell. In addition, supplies of semiconductor grade silicon are running low due to the expanding market for transistors and integrated circuits for computers, calculators, television sets and digital watches, and now the rising demand for photovoltaic systems.

Fortunately, photovoltaic devices do not have to use silicon as pure as semiconductor grade. They can tolerate ten times as much impurity as other semiconductor applications—as much as .000001 percent. New processes have been developed by American firms to make "solar grade" silicon as well as semiconductor grade, and estimated prices of the product range from $10–$20 per kilogram.[1] One or more new plants using these processes are expected to be in operation by 1983–84. The projected decline in the cost of photovoltaic systems is predicated in part on the availability of silicon of adequate purity at a lower price.

The Siemens process has been used for twenty years to provide semiconductor grade silicon for the electronics industry. Very pure compounds of silicon are reacted with an electrically heated seed rod. Ultrapure silicon is slowly deposited on a cherry red rod, which grows into an ingot weighing about 220 lbs (100 kg). This is a high temperature, slow, batch process and the product is expensive, primarily due to the larger amount of heat energy consumed.

Though a solid block, the ingot is not a single crystal but a jumble of tiny crystals known as polycrystalline silicon, or polysilicon.

Base Option: Single Crystal Silicon

The next step is to produce single crystal silicon in block form. This is done by remelting the polysilicon in a special furnace where the temperature is

[1] A listing of companies producing silicon cells is provided in Appendix 4.

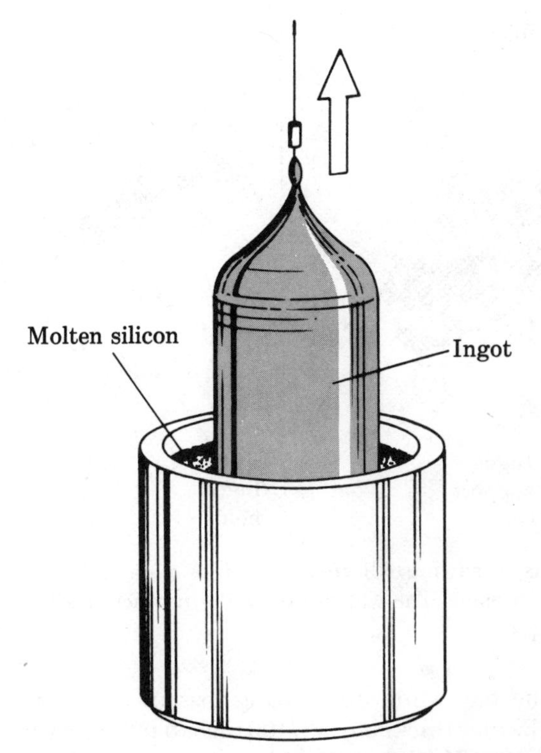

Fig. 3–1 An ingot of single crystal silicon being grown from a crucible of molten material (Czochralski Process). The atoms of silicon solidify in perfect cubic pattern following the structure of a seed crystal that is immersed in the melt, then slowly drawn upward. Under carefully controlled conditions, growth rates of as high as 4 in. (10 cm) per hour are possible.

carefully regulated. In the Czochralski process a seed crystal of silicon on a holder is dipped into the melt and then very slowly drawn upward by mechanical control at a rate of perhaps an inch every three hours (Fig. 3–1). The molten material starts freezing (solidifying) on the seed crystal, following the identical crystalline pattern. In the most advanced production units available, single crystal ingots can be formed 6 inches in diameter and up to 36 inches long. The price of crystal ingot is $100–$200 per pound. Monsanto Crystal Systems, Silicon Technology, Siltec, and Varian are leading producers of single crystal ingots. To reduce costs, development effort is underway on continuously fed, higher speed, pulling operations.

After cooling, the cylinder is laid on its side and carefully sawed into thin sheets or wafers (Fig. 3–2). The object is to obtain as many wafers as

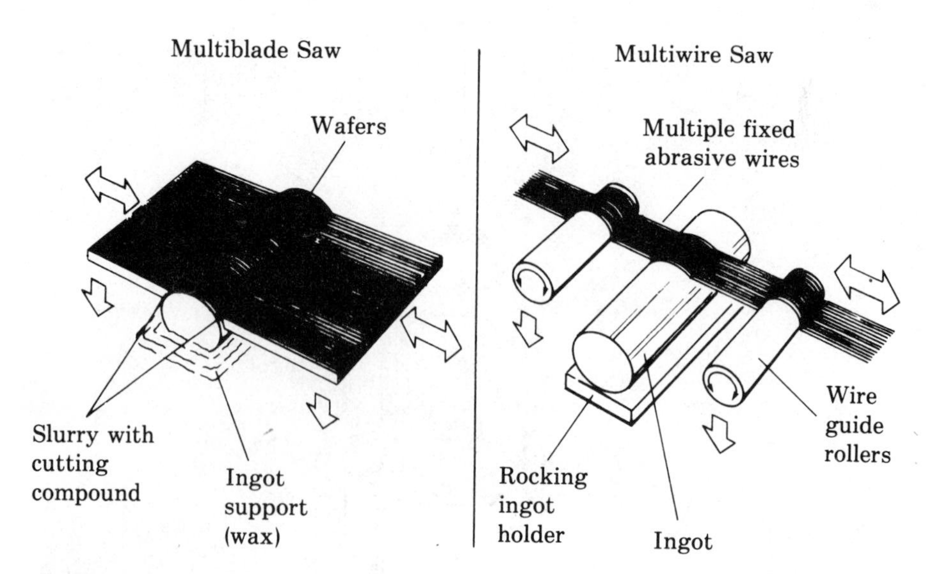

Fig. 3–2 Methods of sawing silicon crystal into wafers. The cutting surfaces contain diamond abrasive. The wafers are 1/100-in. thick. Half of the expensive silicon bar is lost as dust.

possible from one bar while minimizing loss in the form of "kerf," or sawdust. Typical wafer thickness is 1/100 inch (0.025 cm); while kerf losses are about 50 percent. Much has been done to improve sawing techniques and speeds. Thinner saws, multibladed circular saws, and wire saws of various descriptions have been developed, some of which can cut 1,000 wafers at a time. Companies prominent in wafering are Hamco, Siltec, and Varian.

Doping of the slice to create the two layers required for the photovoltaic phenomenon is accomplished in two steps. A trace of boron (or another element such as gallium) is first added to the molten silicon and thoroughly blended in. The ingot, as grown, is thus all p-silicon. After the wafers have been cut, they are coated on one side, then passed through a furnace containing a vapor of phosphorus (or another element with five outer shell electrons) for just enough time to allow the phosphorus to diffuse a short distance into the silicon, converting it into a very thin layer of n-silicon. At the zone of deepest penetration of the phosphorus, the barrier of static electricity spontaneously and immediately establishes itself. Metal contacts are then applied to each side—usually a metal plate to the bottom or p-side, and a grid of metallic fingers to the top n-side. Finally, the entire cell is encapsulated in a weather-resistant, transparent coating, such as a silicone plastic (Fig. 3–3).

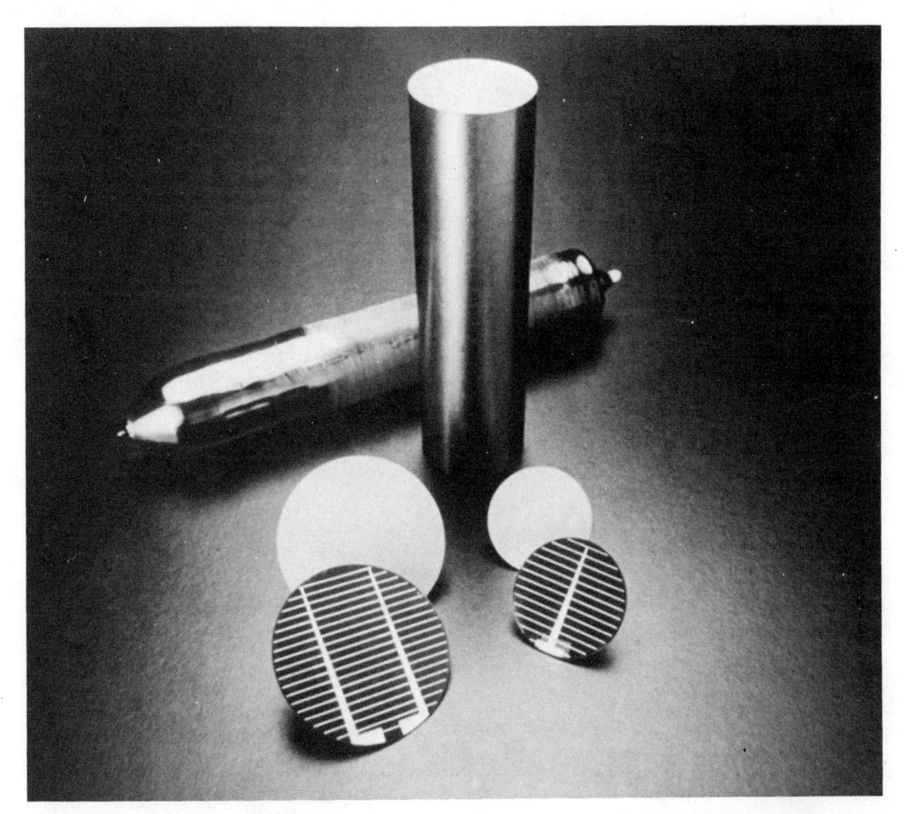

Fig. 3–3 Czochralski ingots, wafers and cells.

If the pure silicon can be supplied for this base process at $15 per kilogram, if efficiencies are increased from 10 to 16–18 percent, if continuous, high-speed pullers are developed, if high-speed, multiple blade slicing is achieved, and if automated production is implemented, single crystal silicon arrays can be sold profitably at prices less than $1 per peak Watt. At this array price photovoltaic systems can be made and installed at costs of $2–$3 per peak Watt, generating electricity that is cost-competitive (four to twelve cents per kilowatt hour) with alternative supplies in many areas. Over 100 industry and university teams are researching these approaches to cost reduction.

Option 2: Silicon Sheet

To avoid the two expensive steps in making single-crystal wafers—pulling the bar and slicing it—methods have been developed to produce a sheet of silicon that is nearly single-crystalline in structure. There are two major sheet processes, the ribbon and dendrite processes.

In the ribbon process, a graphite or ceramic die, or shaping guide, that has a thin slit through the center is lowered into a pool of molten silicon. The silicon is drawn up through the slit by capillary action. A seed crystal is touched to the melt at the top of the die and slowly withdrawn upward. As the silicon leaves the die, it slowly cools into pure crystalline form following the single crystal pattern established by the seed. The melt is continuously replaced in this and similar sheet processes. This process is also called edge-defined film-fed growth (EFG for short). The product is a ribbon 4–5 inches wide (10–13 cm) and less than 1/100 of an inch thick, which is wound onto a drum above the melt. Subsequently, it is cut into sheets of suitable length and doped and surfaced just as the single crystal wafers were. The resulting cell is rectangular, preferable to the circular wafer.

The key companies involved in ribbon growth are Mobil-Tyco, Solar Energy Corp., Energy Materials Corp., Motorola, and IBM. The quality of the silicon produced by this method is adequate to produce 10–12 percent efficient rectangular cells that can be combined to make 8–10 percent efficient arrays. If one does a detailed cost analysis of this large production volume, low-material-consuming option, it can be shown that array prices below $1 per peak Watt can be obtained.

The dendrite process developed by Westinghouse does not use a shaping guide. Very simply, two special seed crystals, called dendrites, are

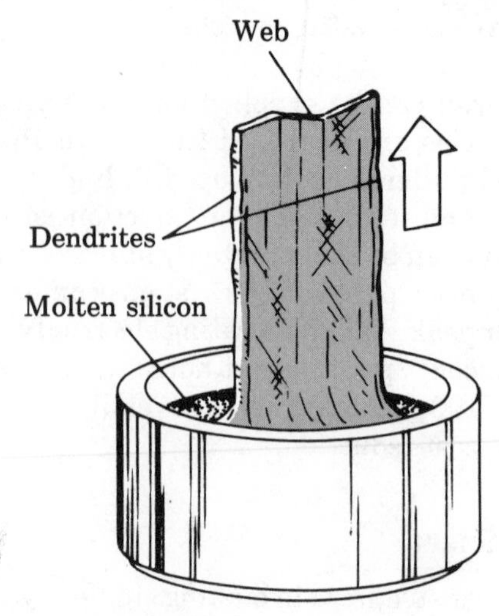

Fig. 3–4 Web dendritic sheet growth.

Fig. 3–5 A strip of crystalline silicon ribbon (lower left) grown by the dendritic web method, and a module (right) composed of cells made from this material. The ribbon is 1.6 in. (4 cm) wide and 0.012 in. (300 micrometers) thick.

dipped side by side into molten silicon. The crystals are pulled out and up from the melt and a sheet of excellent quality silicon crystal is grown continuously without slicing (Fig. 3–4). The solar cells made from dendrite sheets are excellent, with 16 percent efficiency the rule (Fig. 3–5).

The technical feasibility of both the ribbon and the dendritic web processes has been established. Using experimental equipment, the pull rates, sheet width and thickness, quality, continuous processing and material consumption that are required to reduce the cost of arrays by the necessary factor of 10 have been demonstrated. It is possible that within two years factories making continuous silicon sheet could be built, assuring array costs and prices even lower than those forecast for the basic single crystal sliced product (Fig. 3–6).

Option 3: Ingot Casting of Nearly Single Crystals (Polycrystalline Silicon)

Single crystal pulling can be replaced by a casting process that produces predominantly single crystal material, a multicrystal ingot, which is then sliced using the advanced slicing technology of the Base Option. Pure molten silicon at about 2,000°F (1,100°C) is poured into a special pot or crucible and allowed to solidify in a controlled manner so that a large, nearly perfect single crystal is obtained. Alternatively, the molten silicon is allowed to "freeze" under slightly different conditions yielding a single

Fig. 3–6 Future silicon sheet production. In this artist's conception of a large-scale production unit for silicon sheet, the material crystallizes as it is slowly withdrawn from molten baths. In practice, this would be done in an enclosed unit under an inert atmosphere and carefully controlled conditions.

crystal block of material (Fig. 3–7). Under development by Crystal Systems Inc., this process is called the Heat Exchange Method (HEM). A different process resulting in large crystal polysilicon cubes has been pioneered on a proprietary basis by a subsidiary of Solarex Corp., called Semix for semicrystalline material. A German company, Wacker Chemie, is also pursuing the multicrystalline option. The two processes are compared in Table 3–1.

Recent comparison indicates that the Semix process is faster and uses much less energy than the HEM process. Both processes reportedly can use less pure silicon than the Czochralski method. The Semix process is claimed to be able to use silicon of much lower purity than the HEM process uses. The single crystal advocates assert that semicrystalline material will be harder to saw because of chipping, and that device efficiency yields will be much less than from the single crystal HEM material. The products of both these processes were sold as arrays in 1980. Cost analyses of HEM and semicrystalline material show that array price goals

Fig. 3–7 Square ingot by the heat exchanger method. More than 6 in. on a side, this single-crystal block of silicon weighing 21 lb (9.5 kg) was produced by slow cooling on a seed crystal. After sawing, the wafers will be made into cells and assembled into a compact module. The advantage of the rectangular shape is that the wafers, after slicing, make square cells that can be fitted closely together in a module, saving space without wasting silicon.

of less than $1 per peak Watt can, indeed, be met by either method if pure silicon is available at $15/kg. If claims that HEM and Semix use less pure material are substantiated, the requisite low array prices will clearly be achievable.

Table 3–1. HEM *vs* Semix

Property	*HEM*	*Semix*
Feasibility	Yes	Yes
Material	Single crystal	Multicrystal
Device Quality	Excellent	Excellent
Cost	Less than crystal pulling	Much less than pulling
Factory Status	Experimental	New proprietary factory

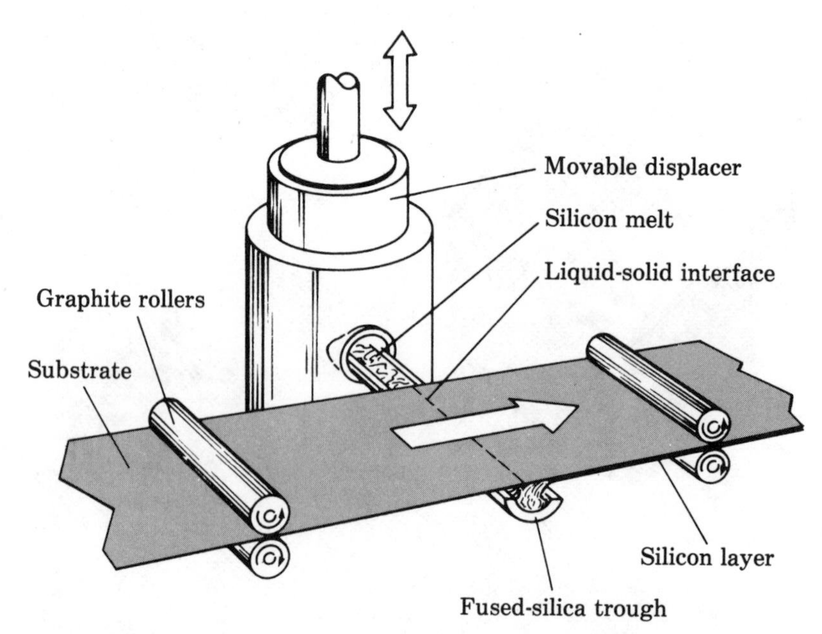

Fig. 3–8 Silicon on ceramic. As a sheet of inert substrate is drawn across a trough of molten silicon, a thin layer of silicon adheres to the underside where it crystallizes under controlled conditions. In the final product, the substrate serves as the base of the cell.

Option 4: Silicon on Ceramics

The farthest departure from the single crystal Czochralski method of producing silicon-based cells is continuous dip coating of silicon on ceramic. This paper-coating-like process continuously creates thin films of silicon on an inert base or substrate which can subsequently be made into modules (Figs. 3–8 and 3–9). The developer, Honeywell, is seriously pursuing the commercialization of the silicon-on-ceramic, or SOC, method. The quality and reliability of film modules have yet to be demonstrated, but there is every reason to believe that a cost-effective product will be developed. Although efficiencies are still below 10 percent, the very thin film requires only one-fourth the silicon required in the Czochralski method, uses much less energy, and has a high output per cost of capital equipment.

All of the silicon methods described above use similar processing equipment and packaging techniques, and are susceptible to the same degree of automation. Any one of the methods apparently will be able to provide economically competitive photovoltaic systems in the mid-1980s.

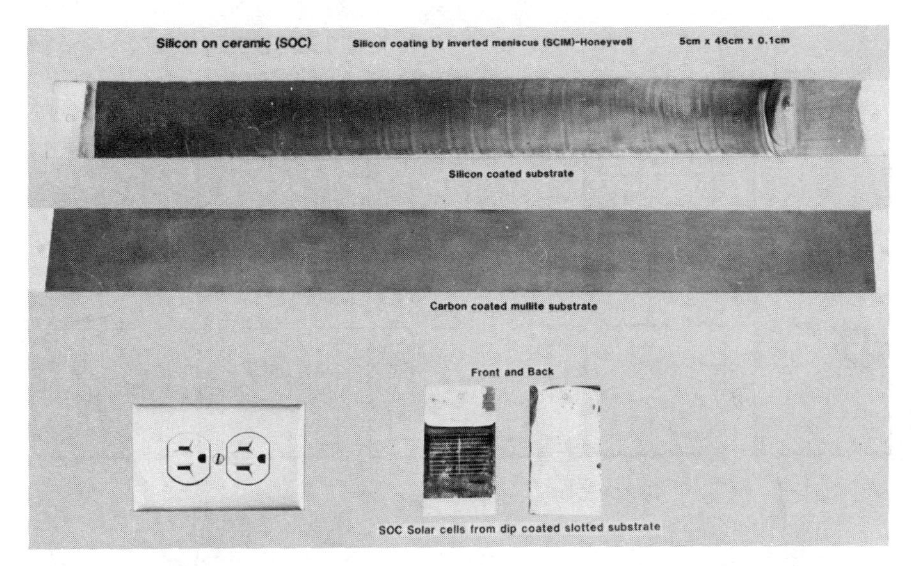

Fig. 3–9 Silicon on ceramic (SOC). The finished strip (top) is 2 in. wide (5 cm) and 0.1 in. thick (0.25 cm).

SOME LONGER SHOTS: MATERIALS AND DESIGNS

Fortunately, but adding to the complexity of the situation for those who must orchestrate the development of photovoltaics, there are several other approaches to designing and producing photovoltaic cells that are equally exciting from the standpoint of reducing costs. In fact, the next three options all appear capable of providing modules that cost less than $.70 per peak Watt.

Option 5: Non-Silicon Thin Films

This category encompasses thin films made with cadmium sulfide, copper sulfide, zinc sulfide, cadmium telluride, and similar compounds. The elegance of devices made from these materials lies in simplicity of manufacture. Some of the materials can be simply spray-coated on glass, dipped in copper, and processed at temperatures less than 400°F (200°C) in a continuous fashion. Because the photosensitive layers are ultrathin, very little material is consumed and cell costs of less than $.20 per peak Watt will soon be possible.

The University of Delaware, in conjunction with DuPont, has developed a detailed plant design, obtaining costs and selling prices much less than $.70 per peak Watt. Figure 3–10 shows an artist's concept of a fully

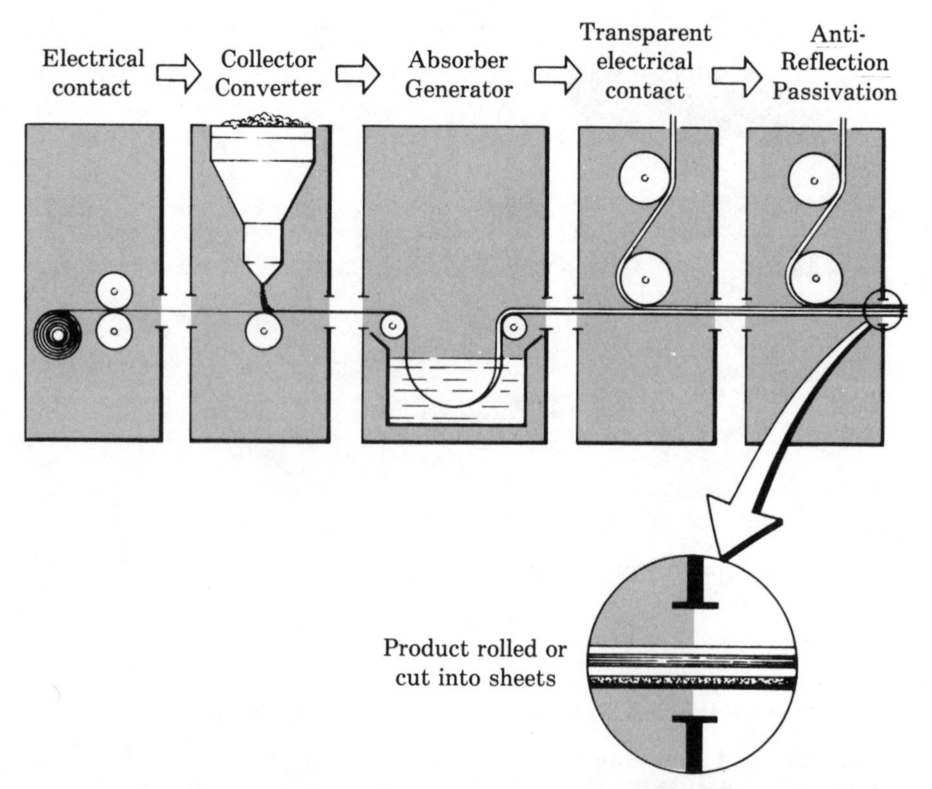

Fig. 3–10 Cadmium sulfide/copper sulfide continuous production process. The schematic diagram above shows how a solar cell might be produced in a continuous process. A roll of thin copper sheet (left) is drawn into the second stage where a thin deposit of cadmium sulfide is added. It then passes through a dip of copper sulfide solution (center), where a thin film plates out to form the cell junction. Next, a thin film of tin oxide, which serves as the upper electrical contact, is added. Finally (right), an anti-reflective coating and weather surface completes the cell structure, which has a total thickness of 2/1,000 in. (50 micrometers). The material has then only to be cut into convenient lengths and electrical connections added.

continuous cadmium sulfide production process, which is not too unlike the production of paper. Photon Power in El Paso, Texas, is presently running a pilot line making 2 ft by 2 ft (61 cm by 61 cm) arrays. Even at efficiencies as low as 4–6 percent, cell costs would be less than $.70 per peak Watt. One might ask, then "If this technology is so inexpensive, why isn't it widely marketed now?" There are two reasons. The low relative efficiency requires the use of larger array areas, increasing area-related costs such as mountings. In addition, the lifetimes of the arrays have yet to be demonstrated. SES, a subsidiary of Shell Oil Company, has been making arrays for nearly two years and testing them thoroughly. Yet until recently Shell management delayed extensive private sales of their cadmium sulfide arrays because the lifetimes were considered too uncertain for the corporation to put its name on the product.

Photon Power has allowed initial testing of its arrays and claims lifetimes in excess of two years. The firm is now constructing a 5-megawatt plant using 4 percent efficient arrays, with improvements expected to yield 6 percent efficiency.

The University of Delaware, in a more sophisticated, more costly process, has demonstrated over 10 percent efficiency on small cells in the laboratory. Again, the key issues are the lifetime and efficiency of cells produced in a continuous, low cost fabrication process. However, the general view is that the cadmium sulfide-copper sulfide manufacturers have a high probability of achieving arrays that cost $.70 per peak Watt or less in the 1982–1986 time frame. The reliability of these materials will in time be proven and large penetration into the market can be expected. This intrinsically simple manufacturing approach may eventually present strong competition to all-silicon technology, at least for some applications.

Option 6: Concentrator Designs

While the photovoltaic cells described so far are mounted flat plate systems that use sunlight without magnification, another option uses optical concentration systems of lenses or mirrors to focus sunlight on very high efficiency cells made of single crystal silicon or gallium arsenide (Fig. 3–11). The cells get hot, which is a disadvantage because cell efficiencies decline at higher temperatures. However, this disadvantage can be partially offset if the heat is used in addition to the electricity.

Using concentrators made of plastic, glass, and aluminum decreases the need for more expensive semiconductor material. If the sunlight is concentrated 100 times, 1/100 of the expensive semiconductor solar cells is required. The high efficiency of the clever designs developed by the concentrator industry have led to forecasts from the private sector for selling prices as low as $.30 per peak Watt in the mid-1980s. Typical optical configurations for concentrating light are shown in Figure 3–12. Recently the Department of Energy (DOE) announced the award of contracts for five large concentrator experiments to supplement its $10 million concentrator cost reduction effort. Using expensive compound semiconductor cells, Varian Associates reports efficiencies as high as 33 percent for cells and 25 percent for the entire system.

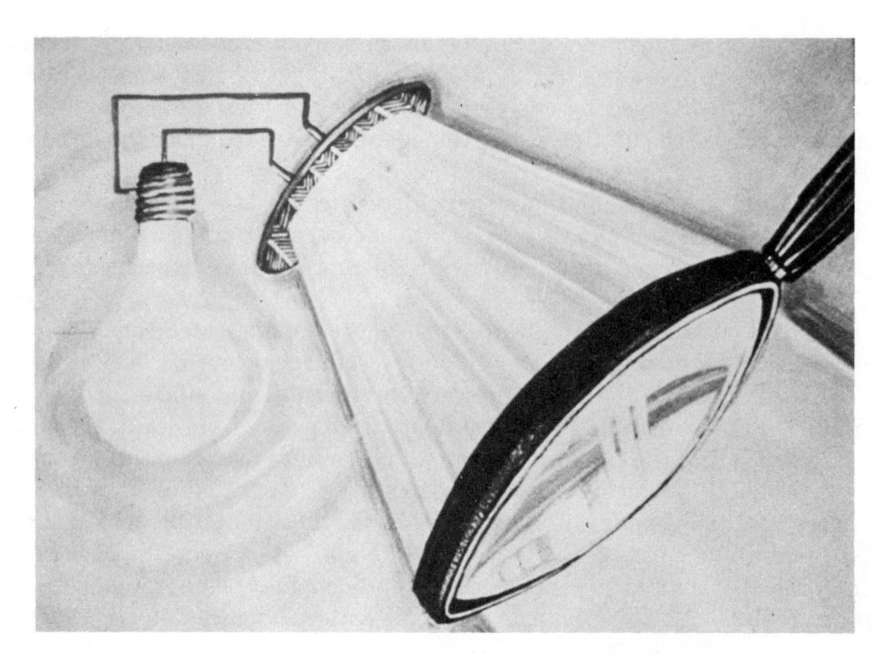

Fig. 3–11 Concentrator cell. By focusing sunlight on a small area, more expensive cells can be used, often with increased overall efficiency. The disadvantages are that the entire unit must track the sun. Diffuse light is not utilized. Concentrator systems are economic primarily in areas that have a large number of clear days during the year.

Four basic approaches are being tried to increase electrical conversion efficiencies of photovoltaic concentrators. The first is to make the standard single silicon and gallium arsenide solar cells themselves more efficient. Recent progress by Microwave Associates indicates that silicon concentrators using sunlight concentrated as much as 1,000 times can convert at 22–24 percent efficiency, based on the incoming sunlight.

Fig. 3–12 Three leading methods of concentrating sunlight on photovoltaic cells. ➧ (a) Parabolic trough reflector focuses sunlight on a string of cells as indicated in the cross-sectional diagram. The reflector is rotated to follow the sun. (b) Linear Fresnel lens arrangement. The lens focuses sunlight on the row of cells just as a large magnifying glass would do. The whole device turns to follow the sun. (c) Point focus Fresnel lens concentrates sunlight on a single cell. This arrangement requires two-axis tracking in order to follow the sun exactly. All require a coolant, usually a liquid pumped across the back side of the cells.

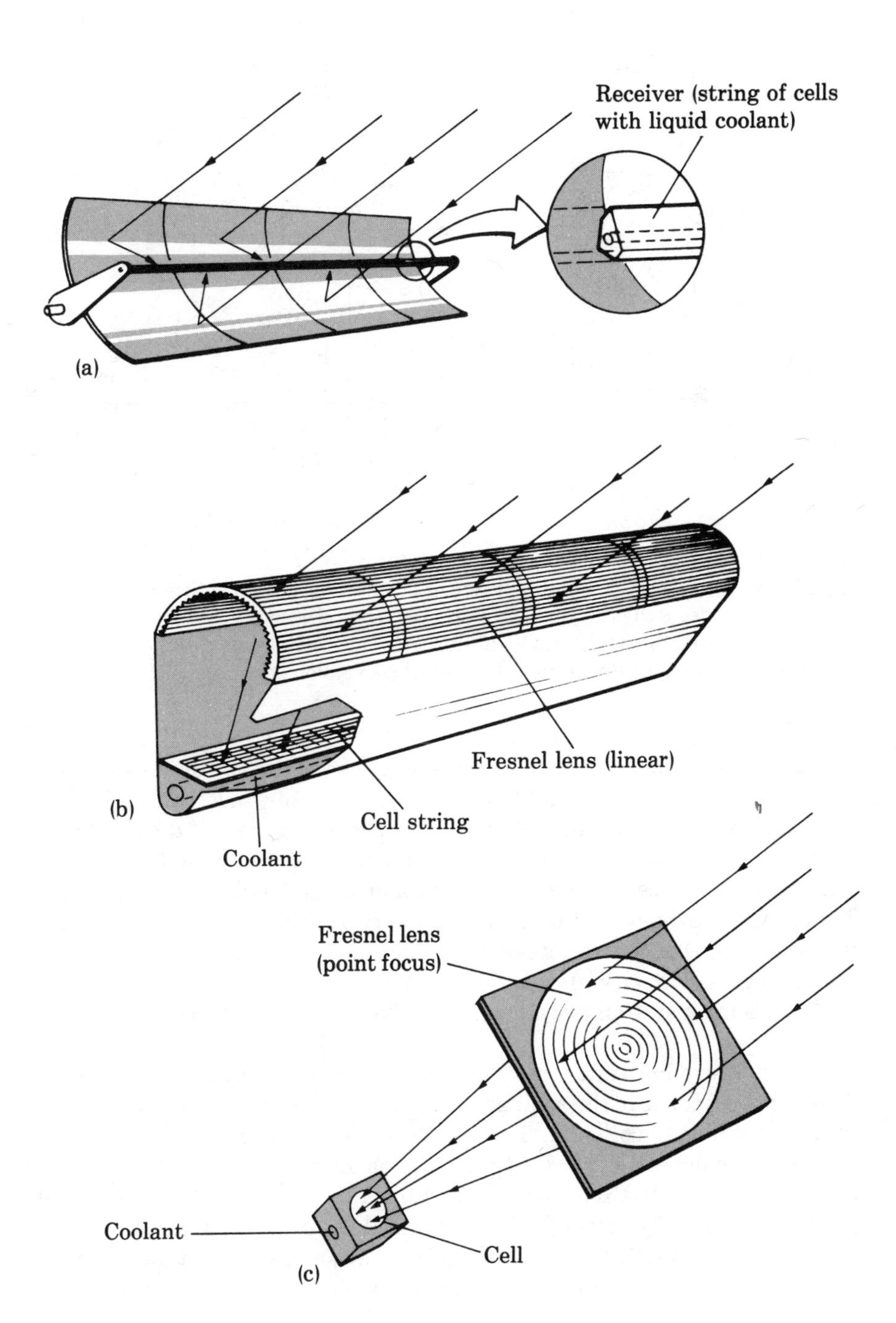
Receiver (string of cells with liquid coolant)
(a)
Fresnel lens (linear)
(b)
Cell string
Coolant
Fresnel lens (point focus)
Coolant
Cell
(c)

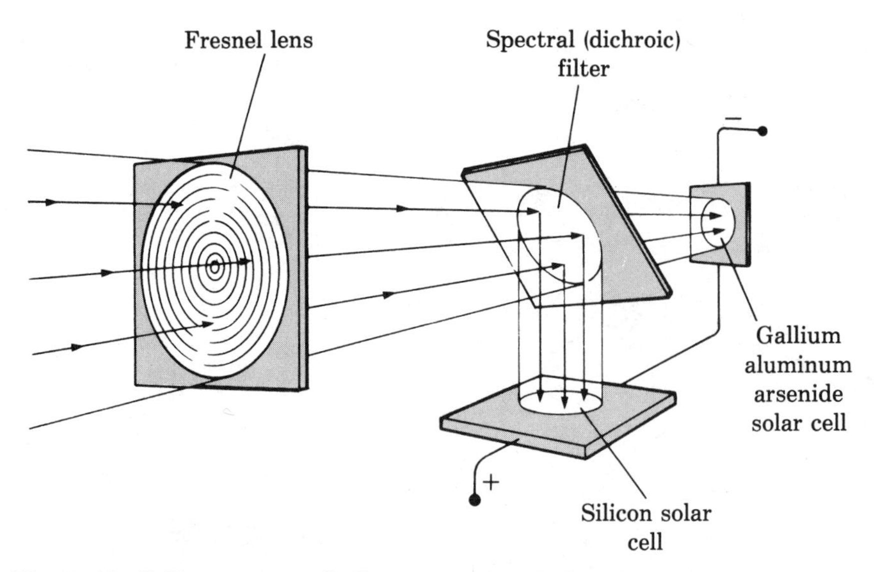

Fig. 3–13 Split spectrum cell. Concentrated sunlight is passed through a special filter which reflects the red part of the spectrum to a silicon cell and allows the blue part to pass through to activate a gallium arsenide cell. Because each cell is more responsive to the light it receives, the overall conversion efficiency is increased.

The second approach is to split the solar beam into two or more parts of different wavelengths (i.e., different colors), each of which is then focused on a separate photovoltaic cell most sensitive to that part of the spectrum. At Varian Associates, laboratory experiments employing dual silicon and gallium arsenide cells have achieved over 30 percent efficiency (Fig. 3–13). There is no reason why this elegant use of two or three cells cannot lead to manufactured concentrator systems which convert 25 percent of the solar energy into electricity and the remaining 75 percent into heat at 212°F (100°C). Such a system is well suited to the industrial electricity and heat market.

In the third high efficiency approach, semiconductors having different frequency response to the solar spectrum are stacked on top of each other to make a composite collector. These cells are still under development, but theoretical calculations indicate conversion efficiencies as high as 60 percent are possible. In this multiple junction or tandem cell system, sunlight passes successively through three thin cells of different materials lying one above the other (Fig. 3–14). The top cell absorbs and is activated by the high energy photons, the next cell converts photons of lesser energy, and

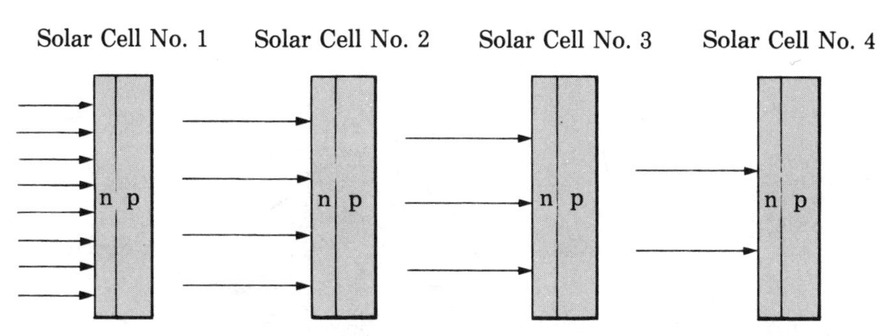

Fig. 3–14 Tandem multiple junction cell. In the multiple junction tandem cell system, three or more cells of different materials are stacked one on top of the other. The top cell absorbs and is activated by the higher energy photons; the next cell down converts photons of lesser energy; and the bottom device absorbs still longer wavelengths. Efficiencies of 50 percent have been achieved in concentrator systems. Even though more complex, such systems may in time become economically practical when very thin films are in mass production.

the bottom device absorbs still longer wavelengths. Efficiencies of 35–40 percent have been achieved. Even though more complex, such systems may in time become economically practical when very thin films are in mass production.

The fourth high efficiency approach is called thermal photovoltaics (TPV). This process, being developed at Stanford University, uses concentrated solar energy to heat a radiator surface to a very high temperature. This radiator then emits the energy in a different part of the spectrum which better matches the silicon target cell. This concept is depicted in Figure 3–15. The silicon cell converts the reradiated energy to electricity. Recent calculations by SAI Corporation for the Electric Power Research Institute indicate 30–35 percent overall conversion efficiencies can be obtained using the TPV process.

It seems clear that high efficiency photovoltaic concentrators will find a niche in the marketplace. The wide range of concentrator configurations, cell approaches, and materials of fabrication all forecast a period of intense competition, as well as some confusion, as the various options sort themselves out in the marketplace.

Option 7: Amorphous Materials

The ultimate solution for photovoltaics could well be the use of amorphous materials deposited on steel or plastic. Very little semiconductor material

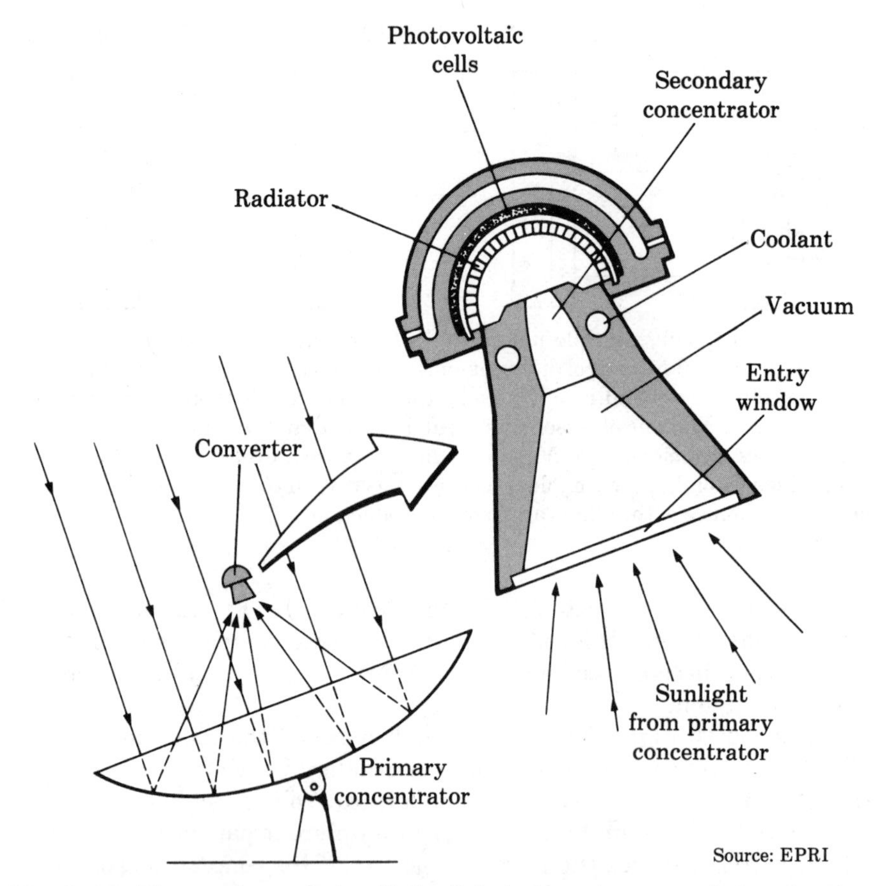

Fig. 3–15 Thermophotovoltaic cell. Sunlight is focused on a metal plate, heating it red hot. The plate then reradiates the heat at a longer wavelength, and the latter activates the photovoltaic cell. The silicon cell is more sensitive to the longer wavelength, increasing overall conversion efficiency.

is used, expensive crystallization procedures are bypassed, and the amorphous state is very durable. Presently, 6 percent efficient amorphous silicon films doped with hydrogen are being produced by Radio Corporation of America (RCA). Other developers have taken the amorphous silicon and added fluorine instead of hydrogen to make improved material.

Energy Conversion Devices, Inc. (ECD) has shown in theory that fluorine-silicon thin film can provide 10 percent efficiencies at quite low cost. ECD is likely to make cells with higher efficiencies than the 3–4 percent efficient cells being sold today. If using amorphous material can provide cell efficiencies of 6 percent, with cell lifetimes of ten years, there is no doubt that selling prices of less than $.70 per peak Watt will be forthcoming, possibly well before 1985–86. Both RCA and ECD have produced experimental cells of about 1 square centimeter with efficiencies greater than 6 percent. We forecast inexpensive commercial cells of 8 percent efficiency by mid-1982.

Several million dollars of risk capital from U.S. industry, in addition to five million dollars per year in federal contracts, have been committed to the investigation of amorphous silicon. The Japanese are also concentrating on amorphous silicon. Recently Sanyo, a Japanese firm, announced a complete line of watches, clocks, and calculators powered by amorphous silicon. Sanyo discovered that amorphous silicon will convert artificial fluorescent light at 6–8 percent efficiency. The low cost of amorphous material coupled with efficiencies close to that of single crystal silicon cells led Sanyo to commercialize amorphous silicon. This may have been a decisive event in the history of photovoltaics.

EXPERIMENTAL CELL TYPES

There are two special experimental cell types that deserve brief mention. The vertical multiple junction cell has tiny grooves cut across the back. Each groove is coated on the inside to form the cell junction, greatly enlarging the barrier area and thus increasing cell efficiency. Conversion efficiencies of about 20 percent have been achieved with this unusual design.

Another unique type is the electrochemical cell (Fig. 3–16). Its special feature is that one layer is liquid. Because the cell junction is formed at the interface between a solution of an electrolyte and a solid semiconductor, some aspects of fabrication are simplified. This cell opens the door to unique applications. For example, the electric energy generated by the cell can be withdrawn for external use as with any other photovoltaic cell, or it may be used within the cell itself to produce useful by-products. In the latter case, materials rather than electricity are withdrawn. If hydrogen is produced, for example, it can be piped off and stored for later use when the sun is not shining. Vigorous research now underway promises an electrochemical cell by 1982 that is 10 percent efficient in converting sunlight into energy. Current research is focusing on several semiconductor materials in single crystal, semicrystalline, and amorphous form for use in electrochemical cells, and on the effects of using different kinds of electrolytes: aqueous, nonaqueous, molten salt, and even a solid electrolyte.

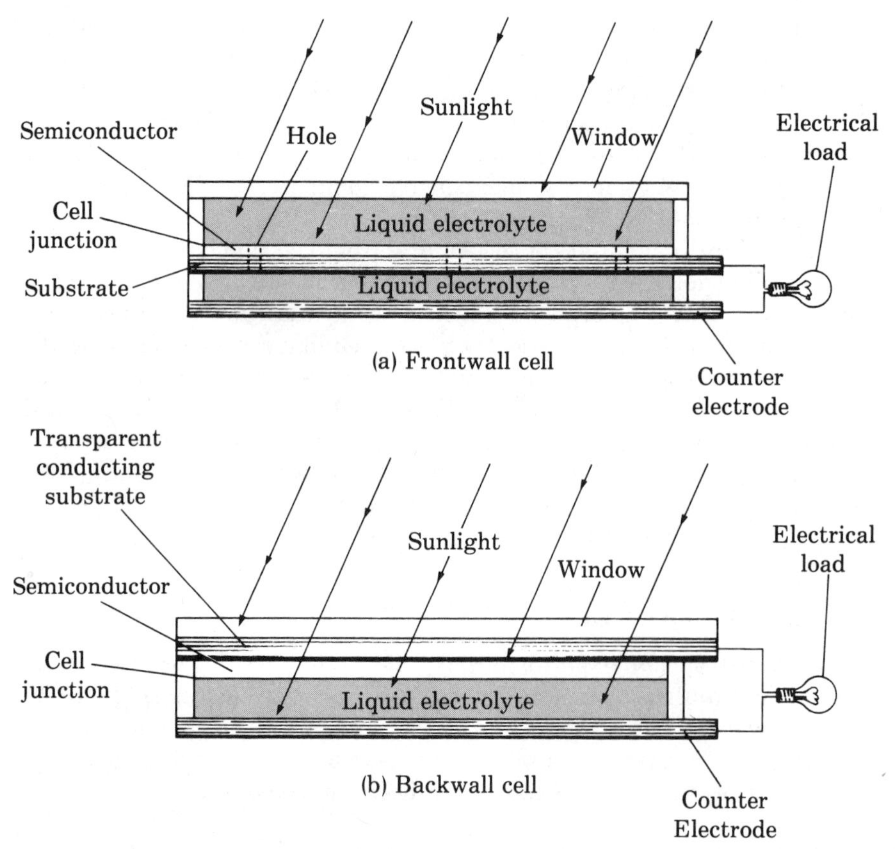

Fig. 3–16 Electrochemical photovoltaic cells. The remarkable versatility of photovoltaics is illustrated by the electrochemical photovoltaic cell, which employs a liquid phase. The cell junction is established at the interface between an electrolyte solution and solid semiconductor material. It was in a simple battery containing a liquid electrolyte that the photovoltaic phenomenon was first observed by Becquerel in 1839. Two designs are shown in which the electrolyte is (a) above and (b) below the cell junction.

A TECHNOLOGY TEEMING WITH POSSIBILITIES

Photovoltaics is a rich and promising technology. The basic photovoltaic phenomenon of converting sunlight to electricity encompasses a diversity of materials and a multiplicity of fabrication techniques, and these can be coupled with a wide assortment of cell designs.

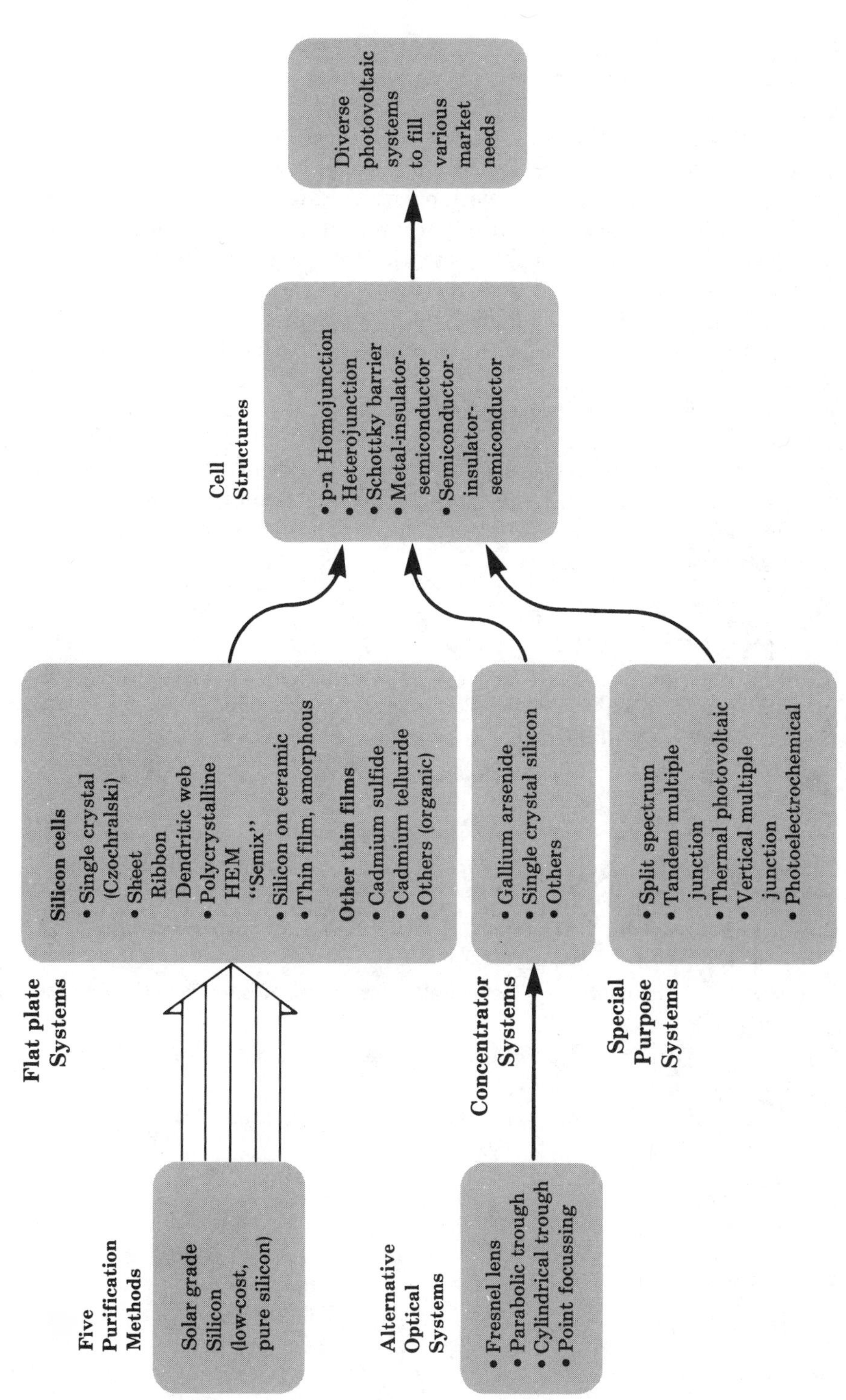

Fig. 3–17 Diversity of approaches to meet various market needs at reduced cost.

To this is added the flexibility of using either flat plate or concentrating optical systems, as well as the potential of concepts such as multiple junction, split spectrum, and thermophotovoltaic cells. Many approaches are available to reduce costs through simplifying and automating production processes. Simultaneously, scientists are learning more about the delicate effects of molecular structure and trace materials on conversion efficiencies.The multiple avenues that are open assure continuing rapid decline in system prices even if there are no further breakthrough discoveries.

The production technology, which is evolving hand-in-hand with advances in cell design, is equally rich in opportunities for discovery and invention. There is, to be sure, an irony in this state of affairs, for the very diversity and promise of the technology diffuses available venture capital among a host of candidate solutions, no one of which is clearly superior or preeminent to all others.

FURTHER READING

Ariotedjo, A. P., and Charles, H. K., Jr. "A Review of Amorphous and Polycrystalline Thin Film Silicon Solar Cell Performance Parameters." *Solar Energy* 24, November 1979, pp. 329–339.

Backus, C. E. "Photovoltaics III: Concentrators." *IEEE Spectrum* 17 (February 1980): 34–36.

Backus, C. E., ed. *Solar Cells.* New York: IEEE Press, 1976.

Carlson, D. E. "Photovoltaics V: Amorphous Silicon Cells." *IEEE Spectrum* 17 (February 1980): 39–41.

Gay, Charles F. "Solar Cell Technology: An Assessment of the State of the Art." *Solar Engineering* 5, March 1980, p. 15.

Loferski, J. J. "Photovoltaics I: Solar-Cell Arrays." *IEEE Spectrum* 17 (February 1980): 26–31.

Loferski, J. J. "Photovoltaics IV: Advanced Materials." *IEEE Spectrum* 17 (February 1980): 37–38.

Wolf, Martin. "Photovoltaics II: Flat Panels." *IEEE Spectrum* 17 (February 1980): 32–33.

CHAPTER 4 Applications

History is responsible for the mistaken and widely held notion that photovoltaics is a high cost, exotic technology with very limited and specialized uses. In the rush to loft the American response to the Soviet Union's Sputnik, Army Signal Corps engineers at Fort Monmouth, N.J., obtained enough design latitude to power the Vanguard I's 5-milliwatt radio transmitter with commercially available solar cells. On March 17, 1958, six small arrays containing 108 silicon chips went into space. But the full implications of using solar power did not sink in until after the launch: with no cutoff device, Vanguard's signals fully and needlessly occupied a radio band for about eight years. The next 22 U.S. satellites went up with electrochemical batteries, but in late 1959 Explorer VI was fitted with 8,000 1- by 2-cm silicon cells, which produced about 15 Watts of power.

Except for a few nuclear reactors, silicon solar cells remained the chosen power source in space. By 1975, the National Aeronautics and Space Administration (NASA) was using nearly a million cells a year. Array sizes grew: Nimbus, a weather satellite launched in August, 1964, carried a 500-Watt array; the Orbiting Astronomical Observatory carried a 1,000-Watt array; some Air Force satellites produced 1.5 kilowatts. Only engineering considerations of balance, rotation, and stress now seem to present limits to the size of photovoltaic arrays in space. A manned space station plan calls for a 100-kilowatt array.

On earth, solar cell production grew rapidly during the early 1960s. Companies such as Hoffman (now Applied Solar Energy), Heliotek, International Rectifier, RCA, and Texas Instruments entered the field. By 1970, only Hoffman and Heliotek remained in the photovoltaics business. Total sales leveled off at about 80 kilowatts per year, at an average cost of $100–200 per Watt.

Success in space led to renewed speculation among photovoltaic producers over potential terrestrial uses. By 1970, concerns about finite fossil fuel supplies and about pollution from excess heat, combustion by-products, and radiation from other energy sources began to place photovoltaic power generation in an ever more favorable light.

In 1971, a NASA employee, the late William Cherry, pointed out that solar arrays covering 31,500 square miles (about 1 percent of the land area in the 48 contiguous states) could supply all the U.S. electric needs then projected for 1990. If the installed price could be reduced to $.50 per Watt, and operate at no better than 7 percent efficiency, farmers could install arrays on otherwise unused land and earn perhaps $2,000 per acre per year from the sale of electric power. Operating costs over a 20-year period would total slightly more than half the costs for an equivalent hydroelectric plant, the next cheapest alternative.

Around 1973, three new manufacturers entered the field with an eye towards a new terrestrial market. Now, about 15 companies are engaged in the manufacture of solar cells in the United States. But all face the common problem of transforming a market perceived as small and specialty-oriented into one characterized by large scale, low unit cost, and widespread general applications.

At the same time that photovoltaics suffers from the misperception that it is an exotic technology with specialized applications only, experience in space and on earth has shown that this technology is well-suited for widespread general use. Solar cells have survived the harsh physical rigors of space—high vacuum, hard radiation, great temperature differentials—and proven themselves durable, efficient, and dependable. In terrestrial applications they have survived dirt, dampness, chemical smog, wind, and hail, and again proven their virtues. Experience has shown that photovoltaic systems can survive in space for twenty years, and that they can survive the terrestrial environment, in many respects even harsher, for at least ten years without signs of fundamental failure. Terrestrial systems have already been built and used for virtually every electrical load, ranging from microwatt systems for digital watches to small remote systems powering houses, villages, and utility substations.

Assuming the ability to produce cheap photovoltaic cells, where and how are these millions and billions of tiny generating units going to be aggregated and used? There are three broad categories for terrestrial photovoltaic applications: stand-alone, residential and other buildings, and central utility generation. Each of these must meet certain common design criteria, yet each has different operating criteria and economies.

One set of design criteria centers on photovoltaics' ability to tolerate the insults heaped on installations by the earth's weather. Many installations over many years have shown that weather, except to the extent that it limits insolation, is not an insurmountable obstacle to photovoltaic deployment in most areas of the world.

It is obvious that outdoor photovoltaic installations must be protected against moisture. This is a simple design problem. Beyond this, any precipitation that covers or condenses on the array surfaces—rain, sleet, snow, fog—can influence how well a solar array performs. Freezing rain or frost both can cause some loss of power until they melt or evaporate with rising temperatures. Worse is snowfall followed by several days of subfreezing temperatures. Until most of the snow has melted from the array surface, little or no power can be generated. Experience with installations at General Electric's Valley Forge, Pa., office building, and the Beverly, Mass. High School, which endure both inclement and variable winter weather, has shown that neither snow cover nor frost has been a significant problem.

If any portion of the array is exposed to sunlight, even an edge or corner of a panel, the heat generated in that small part spreads across the array, which clears itself in a short time. The snow cover slides off the tilted surface, provided there is room at the bottom of the panel. In short, we are not facing a limitation on photovoltaic installations imposed by weather patterns, but a straightforward design problem.

Because ambient air temperature can affect cell temperatures, it affects the efficiency of the photovoltaic generators. The power output of a photovoltaic cell at a given level of insolation is inversely proportional to the cell temperature. For silicon cells, the power output falls about 10 percent for each 45°F (25°C) rise in cell temperature.

It is true that if the cell itself cannot meet design criteria, the system cannot function. However, while the photovoltaic cell is the heart of an installation, it is not the only necessary part (Fig. 4–1). Solar energy (sunshine) is diffuse. The problem this presents to the systems designer is twofold: the system developed must be as efficient as possible in collecting and converting the energy into usable form, and the collectors must cover large areas at minimum cost.

A photovoltaic system must have a structure to mount the cells on and wires to interconnect them and carry off the electricity (Fig. 4–2). Other necessary components of the system depend on how the system is to be used. If power is required when the sun is not shining, a battery for storage or another source of electricity must be included. When storage is used, a special kind of switch is added to prevent electricity from flowing from the battery back into the cells when they are not producing.

Power produced by any photovoltaic device is direct current (DC), as the electrons are flowing steadily in one direction only. DC power can be used directly in some cases; but if the electricity is for household use,

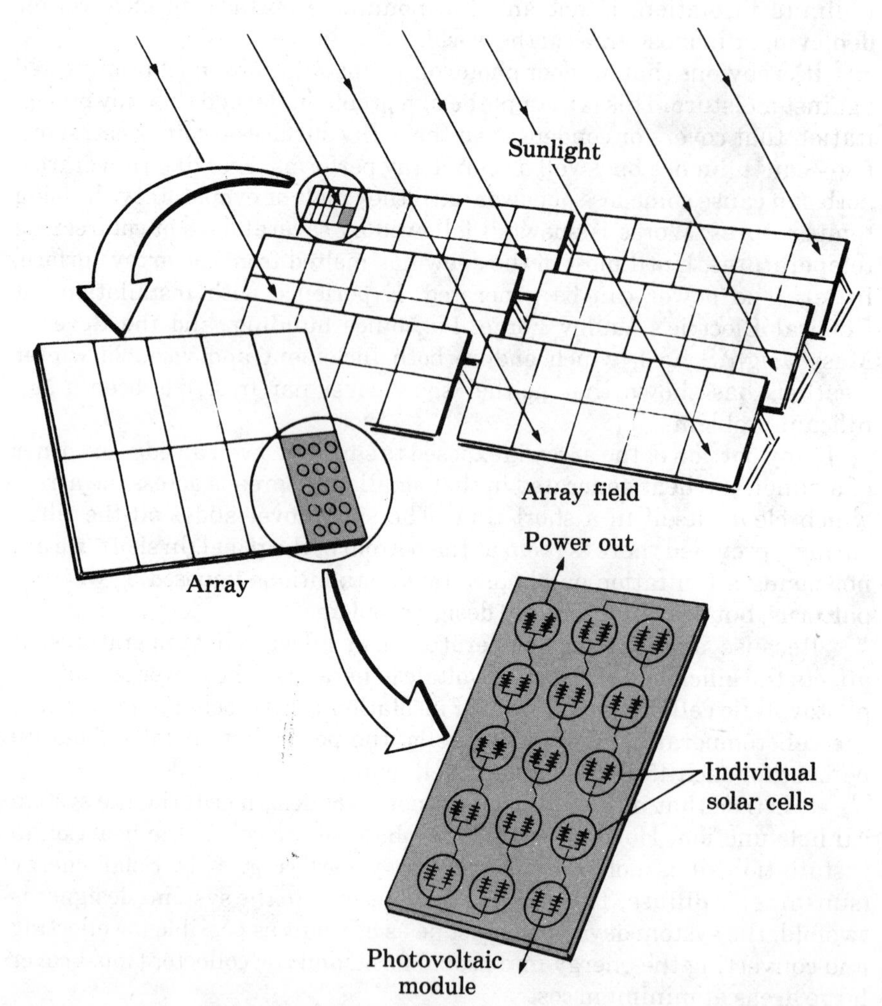

Fig. 4–1 Module. The basic building block for photovoltaic arrays is the module, a sealed panel containing a number of cells interconnected at the factory. It is shipped as a unit, with electrical leads attached for connection to other panels or to an external circuit. Several modules are usually mounted together, forming an array. A collection of arrays is called a field.

alternating current (AC) is required for most appliances. To change DC to AC a power inverter is added. A photovoltaic system may include additional controls to monitor system performance and to regulate power being sold back to the utility.

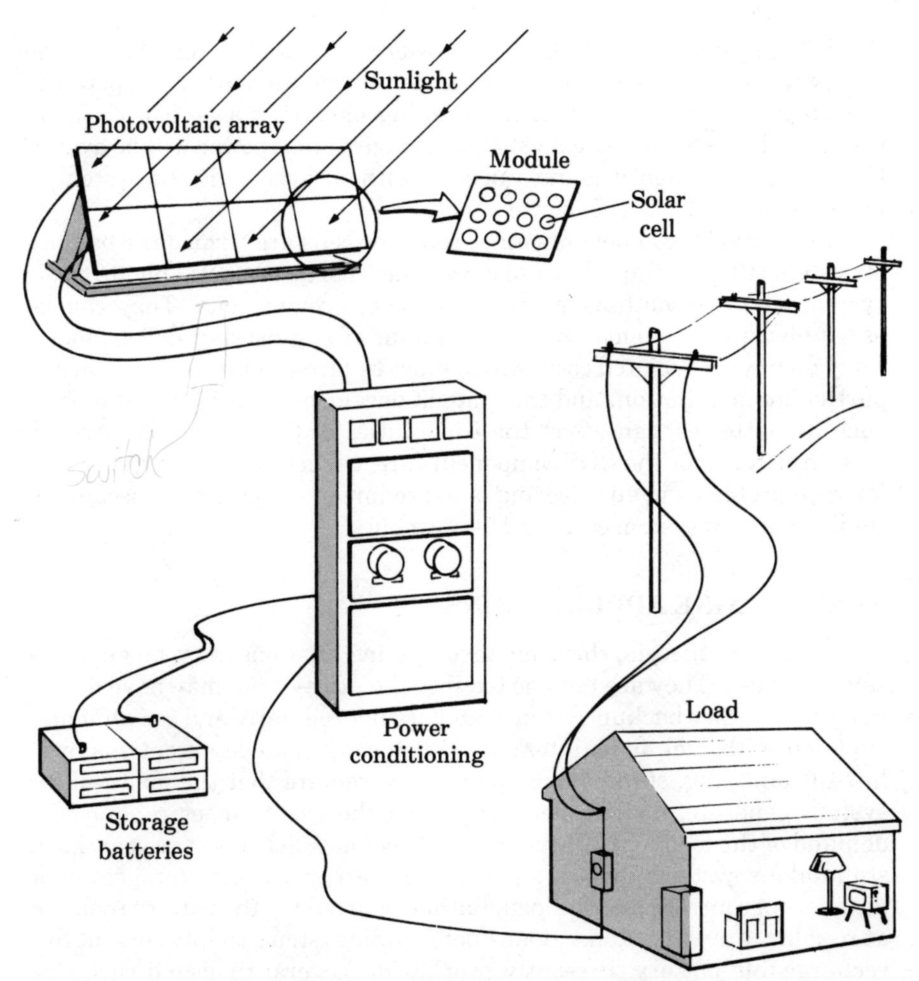

Fig. 4–2 System components. In addition to the array, a photovoltaic system has a power conditioning unit, possibly storage batteries, and a connection to the using equipment or load. It may or may not be interconnected to a utility through a meter.

The components other than the array are sometimes referred to as the balance-of-system, or BOS. There is nothing exotic about these components. They can be assembled into systems with a minimum of engineering development. In mass production, system costs will decline significantly.

If the system "stands alone," to power an isolated village that has no utility service, automatic controls may be incorporated to turn off power to less important equipment in the evening or during a period of cloudy weather. This controlled load shedding assures that power will be available where and when it is most needed without oversizing the system at unnecessary additional expense.

Collectively, components other than the cells are termed the balance of system (BOS). While they are a necessary part of the total photovoltaic system, there is nothing exotic about these components. They can be assembled into systems with a minimum of engineering development. They do have associated costs which must be factored into evaluation of a particular installation, and the state of development of BOS equipment, such as batteries, can affect the capabilities of the system as a whole. Design criteria for the BOS components are, generally speaking, straightforward problems in engineering, and are important primarily because of their impact on system costs and performance.

STAND-ALONE APPLICATIONS

As the name suggests, these photovoltaic installations must be virtually self-sufficient. They are not hooked into the utility grid, may have diesel, battery, or other backup systems, and are not regularly and conveniently supplied with fuel or maintenance. With some notable exceptions, e.g., low-lift pumping, stand-alone applications require that the photovoltaic system generate enough electricity from the sun to meet reliably the demand of the load for twenty-four hours per day, 365 days per year. Most stand-alone systems involve a photovoltaic array, battery storage, minor numbers of controls, and the load. Rather than having the battery replaced at regular intervals, stand-alone photovoltaic systems supply current to a rechargeable battery. Presently, worldwide, several thousand such systems are purchased each year.

Stand-alone systems have proven to be a reliable, very low maintenance method for producing electricity in moderate amounts in isolated unmanned sites. These applications include remote communication repeaters and receivers; remote sensing stations; remote lighting systems; signals, including river and ocean navigational aids; and cathodic protection of remote bridges and pipelines. Such applications are too remote for regular maintenance or for fuel delivery. Experience with a variety of stand-alone systems has been excellent, with virtually no failures in all types of weather.

The hand-held calculator is an interesting example which illustrates the potential of portable "stand-alone" systems. Several models (Sharp and Teal) now on the market are fully powered by the light of incandescent and fluorescent lights. The photovoltaic generator consists of nine solar cells connected in series. Each cell is about 1/4 inch by 1/2 inch and the entire array is 1/2 inch by 2 inches. This array generates a few thousandths of a Watt at 3–4 volts. There is no storage in the system, nor is there an on-off switch. When the calculator cover is opened, the light hits the solar cell, generates electricity, and the calculator is ready for use. After calculation is complete, the cover is folded over the cells, the generation ceases, and the system is off. This light-powered calculator is available for less than $40 retail. No batteries are ever required.

In the class of stand-alone photovoltaic systems there are many portable, low power (less than 100 Watts), lightweight applications. The list of portable photovoltaic systems currently available includes photovoltaic-powered digital watches, clocks, radios, televisions and other portable electronic devices. More rugged applications include portable systems for charging the batteries of fishing and sailboats, caboose lights for trains, and military portable communications. All of these applications are now commercially available and economically sound. These systems supply power to their loads or alternatively provide trickle-charging for batteries to assure that electrical equipment always has adequate stored reserve, even when the system is infrequently activated or visited.

Reliability and readiness are especially important in military applications. There are many situations in which a reliable, portable source of electricity is required. Figures 4–3 and 4–4 show two very different portable applications. One is a lightweight photovoltaic panel that charges rechargeable batteries for a portable field telephone system. Hundreds of these systems are being tested throughout our military. The other shown is a truck-transported mobile communications system. Vehicle-mounted photovoltaic systems can generate as much as a peak kilowatt from deployable fold-out panels. Experience to date has shown them to be safe and dependable.

Another excellent example of portable remote photovoltaic power is shown in Figure 4–5. The recent requirement that a railway caboose have a light presented a difficult, costly decision. Southern Railway chose to try a system built around a photovoltaic generator, battery storage, and a direct current light. The photovoltaic array is less than one meter square, and generates less than 100 Watts of electricity in a bright noon-time sun. All of Southern Railway's local service cabooses are being converted to this solar powered lighting. At first thought, a nighttime load for a daytime source could appear to be an odd choice. However, the photovoltaic system works well, has low maintenance, consumes no fuel, and is economic now for such small, portable electrical applications.

Fig. 4–3 Radio relay equipment with solar cell module.

Fig. 4–4 Van-mounted solar arrays used in the Department of Defense's telephone central van experiment.

Fig. 4–5 Solar-lighted caboose. What's the cheapest way to light a caboose at night? With sunlight, of course. A square yard of photovoltaic array and a battery does the job for Southern Railway.

Another major set of applications for the stand-alone photovoltaic systems are the fixed, large power systems. These systems presently include remote forest and park ranger sites, remote military bases, island locations, isolated residences, and remote villages. Today these remote, large power systems are being built as experiments to prove that large systems can function with little maintenance, be operated by semi-skilled persons, and deliver electricity throughout the year with high reliability.

The largest stand-alone system operating today is at the Natural Bridges National Monument Park in Utah (Fig. 4–6). This 100-kilowatt (kW) system is automatically backed up by a diesel generator and provides all the electricity for a small community of Park Rangers, park maintenance workers, and the visitor center. Almost fully automatic, it is the pioneer for many small community systems throughout the world. The basic design has been improved in a recently announced 500-kW system, to be built using an advanced concentrator photovoltaic array in Saudi Arabia.

Another large photovoltaic unit is the 60-kW system which augments a diesel-electric plant at an Air Force radar station at Mount Laguna,

Fig. 4–6 Natural Bridges National Monument in southeastern Utah. The 100-kW photovoltaic array is located near the visitors' center on a 1.3-acre array plot characterized by a gentle slope to the south and an irregular shape to accommodate rock outcroppings. The variegated appearance of the array is the result of incorporating solar cell modules from three manufacturers. Toward the upper left of the picture can be seen the residence and maintenance areas. Within the maintenance area are the generator and the photovoltaic storage and control buildings.

California, near San Diego (Fig. 4–7). This system could be a precursor for island , village, and community power systems throughout the world. The U.S. military is also using photovoltaic systems for refrigeration, signals, remote communication, remote target systems, and weapon systems power. The MX missile system will use photovoltaics and other sources of renewable energy.

A photovoltaic-powered Indian Village project at Schuchuli, Arizona is perhaps the most exciting of Federal experiments to date (Fig. 4–8). A small, 3.5-kW system (less power than is required to power your home) provides fresh water, lights for 15 homes, a washing machine, lights for the community feast house, a sewing machine and 15 four-cubic-foot refrigerators for 92 people in the village. The village is 17 miles from the nearest electricity. The Indian Health Service estimates that the United States has over 400 remote villages like Schuchuli and 80,000 individual Native American homes without electricity.

The U.S. Native-American experiments are being copied throughout the world. The United Nations has estimated that there are 10 million villages worldwide like Schuchuli, i.e., they have no electricity and no electric utility available. The first international experiment modeled on

Fig. 4–7 Mt. Laguna facility. The 60-kilowatt photovoltaic power generation facility at Mt. Laguna, California, completed in June 1979, was made possible through a cooperative program between the Department of Defense and the Department of Energy.

The primary objective of the Mt. Laguna project is to demonstrate that a solar cell power system without energy storage can effectively augment a remote alternating current (AlCl) power network, providing reliable power with subsequent fuel savings.

Fig. 4–8 Schuchuli Indian village. On December 16, 1978, the Papago Indian village of Schuchuli became the first community in the world to rely entirely on solar energy for its power needs. Schuchuli is located on the western edge of the 2.7-million-acre Papago Indian reservation in southwestern Arizona. The village's 15 families (95 people) live 17 miles from the nearest available electric utility power source, which makes Schuchuli an ideal site for photovoltaics. Before the installation of the photovoltaic system at Schuchuli, a diesel-powered pump provided water, and kerosene lamps and candles provided light.

Schuchuli was funded by AID in Tangaye, Africa (Fig. 4–9). In this application, the photovoltaic system provides fresh water and runs a grain mill to make flour for sale—a major change in the economy of this small community.

In developed and developing countries, one key economic factor in the agricultural application of photovoltaics is the provision of a constant supply to meet electric demands throughout the year. The University of Nebraska, the Kettering Foundation, Lincoln Laboratory of the Massachusetts Institute of Technology (MIT), and DOE have developed a photovoltaic-powered arc process for synthesizing ammonia and making fertilizer. When photovoltaic system prices reach DOE's 1986 target levels, this use will be fully economic. The economics of small, remote applications, such as low-lift pumping, indicate photovoltaics will soon be competitive with today's diesel generators.

Stand-alone remote applications, independent of an electrical grid, now involve relatively small installations far from central power. As costs decrease, photovoltaic systems will in time be built economically to provide virtually all of the electricity the United States requires.

Obviously, if electricity is desired when the sun is not shining, then we must store the solar electricity or have another generator on standby. The most common storage method today is sealed lead-acid batteries like those used on golf carts, industrial forklifts and experimental electric cars. Battery storage works quite well with the direct current photovoltaic array. The cost of storage, however, is another matter. It is about ten times breakeven with power grid costs—lead-acid batteries now cost about $100 per kilowatt hour of storage capacity. They are also limited to about 800–900 refills (or deep discharge cycles). In order to provide genuinely economic storage for photovoltaic systems, the battery industry must reduce cost by a factor of three and increase the discharge life to 2,000–3,000 cycles. Several new kinds of batteries that meet these goals are now under development by the government and industry.

Another storage option is to use photovoltaic electricity to spin a mass at very high speed. This flywheel storage can store enough energy in a 4-foot cube to provide a household with electricity for several days. New magnetic bearings and rotor materials are being developed to reach the cost/performance goals required for flywheels to be an effective storage method for use with photovoltaic systems. The storage industry has confidence that by the mid-1980s, it can reduce the cost of storage so that reliable stand-alone solar photovoltaic systems can be designed and built to provide electricity twenty-four hours a day, day in and day out, for most areas of the United States.

Stand-alone applications, while important for what they can do, do not represent major energy saving opportunities in the United States. Sys-

Fig. 4–9 In the African village of Tangaye, Upper Volta, a photovoltaic power system, a 1.8-kW array, grinds grain and pumps water less expensively than a diesel generator unit. Project sponsored by AID.

tems must be designed and built which connect to and can supplement the trillion dollar electric utility grid which carpets our nation. These larger distributed systems must be long lasting, require little or no maintenance, have minimal impact on our environment, and be fully safe from the user's point of view.

GRID-CONNECTED DISTRIBUTED APPLICATIONS

Grid-connected photovoltaic applications are pivotal; they can save energy, and fulfill many persons' desires to be somewhat independent of the utility and yet enjoy the security of having the grid as a backup.

The primary applications are roof-mounted arrays, facing south and tilted at an angle of latitude plus 10°. The roof array must have sufficient area to generate electricity to serve a large percentage of the building load. Fortunately for most one- and two-story nonindustrial buildings, the south facing roof area receives enough solar energy so the array can generate electricity to serve a large share (greater than 60 percent) of the building's electrical requirements.

Fig. 4–10 Mead, Nebraska facility. An aerial view of the agricultural Flexible Test Facility, Mead, Nebraska. The photovoltaic array field is in the foreground. The experimental unit is operated by the University of Nebraska. The project is funded by the Energy Technology Office of the Department of Energy. The photovoltaic system was designed and constructed by MIT/Lincoln Laboratory.

The most important on-the-grid applications which can generate a large percent of the building load include residences, mobile homes, two-story apartments, low rise offices, small retail shops, schools, colleges, light industry, city and state buildings, barns, and covered parking. Because these applications are not yet fully economical, no private experiments of this size have been built.

DOE, however, has funded several key systems experiments to prove that multikilowatt grid-connected applications can be easily designed, installed, and operated with minimal maintenance. Perhaps the most important on-the-grid experiment completed to date is the 25-kW irrigation experiment at Mead, Nebraska (Fig. 4–10). This first, fully grid-interactive photovoltaic system was fielded by MIT's Lincoln Laboratory. The system has been operating in the extreme Nebraska heat, cold, rain, snow, hail and wind, and has provided reliable electricity for over three years without degradation or extensive maintenance. In fact, the electrical grid that backs up the photovoltaic system has been blacked out three times more often than the photovoltaic experiment. The Nebraska system also shows that photovoltaic power can be converted to alternating current and placed on the power lines without affecting the power grid adversely.

Fig. 4–11 The world's first photovoltaic-powered house under construction in Phoenix, Arizona. The south-facing roof is composed of factory-sealed photovoltaic modules, which fit together with weather-tight connections. This DOE-funded project was built by Long Properties, Phoenix, using a system designed by ARCO Solar Corporation. The system generates 6–8 kW when the solar flux is one kilowatt per square meter. This is enough electricity to cool the house to 78°F when it's 110°F outside, to light a few lights, to cover other normal daytime loads, and still have electricity to sell back to the local utility.

On dedication day, the house actually ran the meter backward while keeping it cool with outside temperatures at 105°F. This particular array design provides the roof for the house as well as the electricity. The total labor to install the first system was only 40 hours. Ultimately, we should see a well designed system arrive at the installation site at 8 a.m. and be powering the air conditioner by noon.

This pioneering system has resulted in DOE funding for several other grid-connected experiments. Most of these are under construction, including the following diverse applications:

A photovoltaic-powered high school in Beverly, Massachusetts
A photovoltaic-powered hospital in Hawaii
An airport power system in Dallas, Texas
The visitor center at Sea World, Florida
A photovoltaic-powered house in Arlington, Texas and another in Phoenix, Arizona (Fig. 4–11), and one near Boston, Massachusetts (Fig. 1–4)

A residential experiment station in Massachusetts and another in New Mexico
Four private homes in Hawaii

How Much Electricity Does Your House Use?

Every utility bill indicates the number of kilowatt hours (kWh) used in the billing period. If a house does not use electricity for heating or air conditioning, it may use 350 to 600 kWh each month, depending on the number of appliances, the number of residents, and the patterns of electric power usage. For space heating and cooling, power usage could be more than twice this amount.

Let us assume that neither heating nor air conditioning is involved and that in a typical month your house uses 420 kWh, or 14 kWh a day.

A median insolation (sunshine) value for the United States is about 1 kilowatt per square meter at noon on a clear day. Considering cloud cover and haze, the total for an average day in, say, Virginia, Missouri, or Northern California is about 5½ kilowatt-hours per square meter. If all this light could be converted into electricity, three square meters of array would provide the power you need.

Because the array is only 15 percent efficient in converting sunlight into electricity, the collector area must be nearly seven times larger, or about 20 square meters in area (about 215 square feet). Most single-family residences could readily accommodate an array of this size or larger on a suitably sloped portion of roof. This means that from the standpoint of size photovoltaic systems are feasible for the single family at present-day photovoltaic efficiences.

Photovoltaics for New Housing

In the case of new housing, adequate roof area usually presents no problem, for it is relatively easy to assure a large, clear south facing roof of proper slope. Indeed, even most multi-family housing can provide this much roof space per dwelling unit. In some cases, two or three times this area could be allocated. This collector area would produce sufficient electricity to operate air conditioning and heating systems, especially if a heat pump is used.

From the point of view of energy efficiency, it is almost unforgivable to see expensive new housing being built that is not oriented to provide a large south facing roof at a good angle to catch the sun's rays.[1] And more often than not, otherwise satisfactory roof areas are chopped up with dormer windows or gables. The nation could ultimately realize enormous energy savings by one simple act: if every zoning jurisdiction in America promptly amended its ordinances to require adequate south facing roofs clear of obstructions and with proper slope, new homes could exploit photovoltaics without major design changes or cost penalties.

By the mid-1980s, solar energy systems will be commercially available that employ the thermal energy produced in the photovoltaic process, as well as the electricity, probably using photovoltaic concentrator designs. Compact dwelling units such as duplexes and multi-unit apartments, where roof area is more limited, will find it advantageous to supply essentially all energy needs by using such total energy systems, even though they may be somewhat more expensive on a cost-per-square-foot basis.

Multi-story apartments and office buildings will make use of south facing walls in addition to the roofs. Photovoltaic thin film collectors will soon be inexpensive enough to make this a desirable strategy. In such cases aesthetics becomes important. Photovoltaic collectors will be available as finished panels, in a variety of sizes and shapes and with different kinds of trim, which can readily be incorporated either as wall or as roof in any new building.

As for existing housing, every case will be different, and ingenuity will be required in providing units for best economic return. A variety of units will soon be on the market with mounts that permit arrays to be adjusted for existing varied roof and wall alignments, slopes, and spaces.

The Cooperating Grid

A complete residential system with 1,000 square feet of array area, generating 3.5 peak kilowatts for new housing in 1986 may cost as little as $7,000. These grid-connected applications have led to serious analysis, technical and political, of the so-called utility interface issue. Can the photovoltaic signal be power conditioned so that the grid electrical quality is not affected? What is a fair price for the utility to charge when it backs up

---

[1] South of an imaginary line across the United States at roughly the Washington, D.C.–St. Louis—Denver—San Francisco latitude, a 45° southward roof slope is about optimum. North of this latitude, the slope southward should be 55° or 60° from the horizontal. The general rule is to add 10° to local latitude to obtain optimum year-round slope southward for a fixed array. For details on siting houses for solar applications, see Sunset Books' excellent *Handbook for Solar Heating—Passive and Active,* or consult a solar architect. An existing roof of any slope can be used if it has a generally southward orientation. Optimum slope becomes less important as the price of photovoltaic arrays continues to decline.

a photovoltaic user? When the photovoltaic system places excess power on the grid, what should the credit from the utility to the photovoltaic user be? Recent regulatory decisions indicate that the person owning a photovoltaic system will find it economically advantageous to have utility backup.

It is unlikely that our society would choose to adjust to a fully distributed power mode where the electricity was out three or four days several times a year, and it is also unlikely that each home would have a diesel generator backing up its photovoltaic system. Also, it is unlikely that the cost of electrochemical storage (e.g., lead-acid batteries) will get to the point where distributed use is more cost-effective than a cooperating utility grid backup. The cooperating grid backup implies a utility network whose pricing structure does not penalize the solar owner, and residential photovoltaic systems that do not impair the functioning of the grid. Many system integration studies by industry have shown that the photovoltaic phenomenon will not affect the grid dynamics, the grid reliability, or the grid reserve capacity until photovoltaic applications supply well above 15 percent of the total demand—and in some grids as high as 25 percent. If photovoltaics supplies more than 15 percent of the demand, the grid will have to obtain additional capacity through either spinning turbine reserve, or additional hydropower for storage, or storage mechanisms, including batteries. The marginal cost of storage is more economic if done at the grid load center rather than on the distributed site.

Given a strategic approach to cooperating with the person who owns a photovoltaic system, the grid can effectively back up the solar option. If grid-connected photovoltaic systems obtain a very high penetration, greater than 15 percent, the utility may have additional costs. If society desires, the utility rate structure can be adjusted to distribute the additional costs to all customers. The value of renewable energy sources to the nation and the worth of solar as an indigenous resource far overshadow the added cost of grid backup. The cooperating grid scenario combines the best attributes of the small-is-beautiful philosophy with the best attributes of Amory Lovins' intense concern for appropriate use at the place of use, while making it possible for industry to continue to grow and make money; it also provides homeowners with the independence that comes from owning their own generation capacity.

The cooperating grid scenario is far from a flight of fancy. In California, Southern California Edison's recent rate demand charge means a person can actually run his meter backwards, use no electricity, and pay only $6 a month to have the backup of the grid assured at his home. A recent law, the Public Utilities Regulatory Policy Act of 1978 (PURPA), requires that a central utility pay for excess power supplied to it by a renewable source such as photovoltaics owned by a customer. The utility m,ust also provide backup power for the solar generator. The exact rates for buying and selling power will be determined by the state utility commissions. PURPA specifies that the buyback rates reflect the "avoided costs" to the utility for not having to generate and transmit the electricity.

New England Electric, in a recent proposal to DOE, has proposed that the marginal cost of fuel be credited for the energy supplied to the grid from a photovoltaic system. The time has begun for a few, very few, select utilities to be pioneers in the cooperating grid scenario which permits the average person to own an electrical and heat generation system, operate and maintain it, gain the economic benefits from solar renewable energy, and at the same time have the reliability of utility grid backup.

If there is a grid failure, another attribute of the cooperating grid scenario is that one still has some solar reserve when the sun shines, so that the entire grid is not knocked out. Given a daytime brownout, one could still keep the refrigerator, freezer, and other critical loads running. The addition of a few hours' worth of storage would provide emergency service in event of a power outage, occasional brownouts, or cloudy days.

CENTRAL UTILITY APPLICATIONS FOR PHOTOVOLTAICS SYSTEMS

Although no photovoltaic experiments of a size required by a central utility have yet been built, all major Sunbelt utilities are participating in large on-the-grid experiments. There appear to be no technical problems associated with multi-megawatt central photovoltaic substations. Moreover, there is adequate land around coal-fired and nuclear reactor utility generating stations, and under transmission rights-of-way, to locate solar photovoltaic systems that could supply a large percentage of central electric power. The Electric Power Research Institute, General Electric, and Westinghouse have extensively studied the technical requirements, the interaction with existing generators, and the economics of this possibility. If the DOE system cost goals of $1,500 per installed kilowatt can be met by 1988–1990, then photovoltaics central power will be fully economic for the Sunbelt in this country. Before this occurs, however, many megawatts of photovoltaic systems probably will have been sold to the distributed users.

It is only appropriate that a review of photovoltaic energy systems conclude with a brief description of a concept that by any measure, be it distance, timing, magnitude or sheer imaginative genius, must be rated the "farthest out"—the satellite power system, or SPS. Huge arrays of photovoltaic cells would be assembled into a large orbiting space platform.

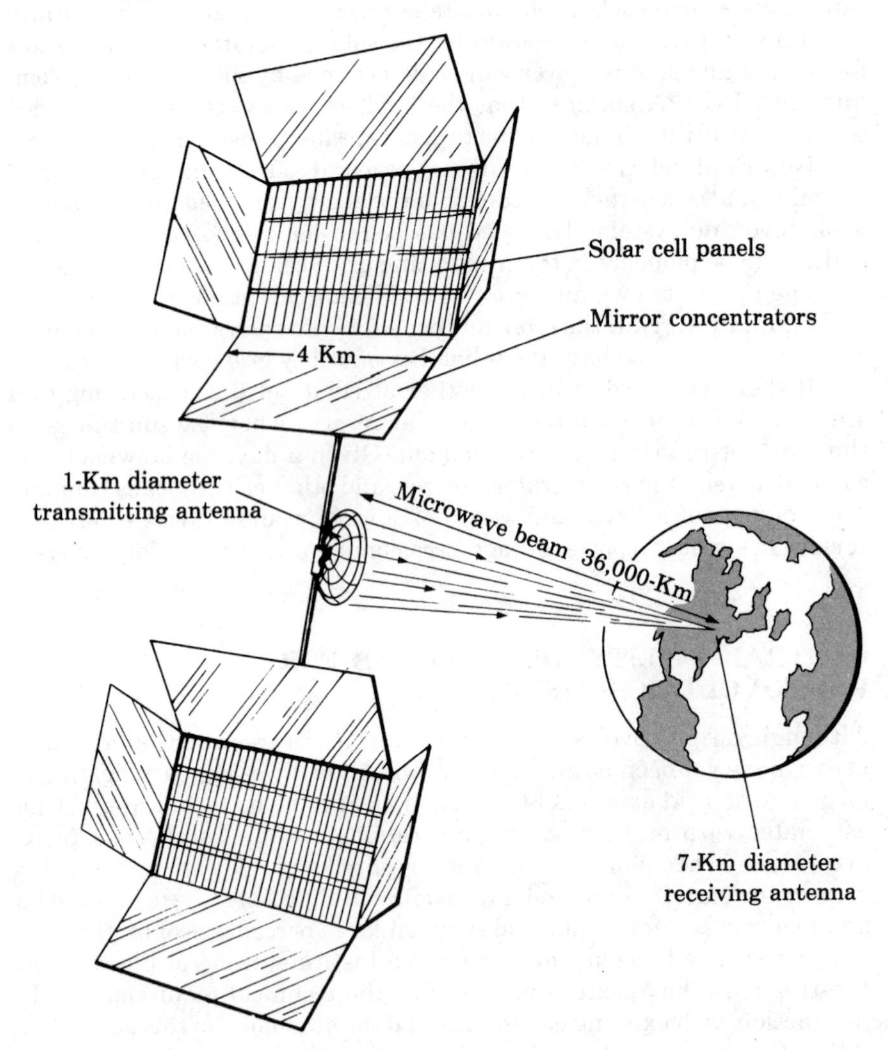

Fig. 4–12 Solar power satellite concept.

There they would collect sunlight and convert it into electricity, unimpeded by weather, day-night cycle of the earth, or seasonal variations of insolation at the earth's surface (Fig. 4–12). As it was produced, the electricity would be converted into microwave energy which would be beamed to earth from the satellite, in geosynchronous orbit at a distance of 22,300 miles (36,000 kilometers).

The array for a single satellite would be some 25 square miles (65 square kilometers) in area, possibly larger, which could deliver to the ground perhaps 10,000 megawatts (10 gigawatts) of electrical energy. The receiving area would occupy about 36 square miles on the ground. People would not live in the receiving area, but a person would not be harmed walking through it. Approximately 25 such satellites could supply all the electricity the United States uses today.

Studies by DOE and NASA indicate that the concept is feasible. The space shuttle would be the workhorse in constructing such a system, and if its troubles are an indication of the magnitude of the technical and economic problems that might be encountered, the SPS may well be relegated to the next century. But the concept has enough merit that a number of these satellites may in time be built to augment earth-based electricity sources. Space stations would be extremely vulnerable to military action, and a relatively peaceful world would be a prerequisite for investing heavily in such systems and then becoming dependent on them. By contrast, highly dispersed photovoltaic systems on the ground, giving each house a source of electricity at least for minimal needs, would be far more secure than the present utility network with its large central station generators.

COMMERCIALIZATION BEGINS

Starting now, photovoltaic systems will find increasingly broad applications in the United States. Recognizing that new applications depend in part on the economic factors affecting the particular site and the owner, the general order in which photovoltaics may appear is suggested by the following list (most apt candidates first):

- Remote villages and houses, existing and new
- Sunbelt residences with utility service
- Remote water pumps for irrigation
- Existing residences in more northerly states
- Educational institutions
- Residential condominiums, high density housing
- Small municipally owned utility systems
- Municipal buildings, hospitals, and other public and quasi-publicly owned structures.

These early markets will enjoy explosive growth in the mid-1980s leading to a self-sufficient, billion-dollar industry by 1986. As costs drop, the latter

1980s will see increasing economic viability for photovoltaics in light industrial complexes, and utility installations near major load centers.

Because photovoltaic systems produce direct current, applications that can use DC power are favored candidates especially if battery storage systems are already in place or if an intermittent power supply is acceptable. Such applications include water pumping for irrigation (no batteries); remote cooling, in which the energy produced is stored in the ice; water desalination; industrial processes such as the manufacture of fertilizers and some chemicals (no batteries); and highway lighting (with batteries).

As photovoltaic technology permeates society, the distinctions between particular applications will become increasingly blurred. Continued expansion will become more a function of proven economic benefits, of the reliability and convenience of photovoltaic systems, and of personal preference for the greater feeling of independence that comes from owning one's power supply.

DOE is supporting an extensive series of test applications of photovoltaic systems of various sizes, for a variety of uses, and under widely differing regional conditions, both as a means of perfecting the technology and in order to give the public an opportunity to observe photovoltaics in use. Some of these projects are identified in Appendix 2.

A field assessment of factors that motivated homeowners in one area to procure solar hot water heaters, while not directly applicable, is indicative of factors that may affect photovoltaics:

- Existence of a local ordinance exempting the owner from property tax on the solar equipment;
- Dwelling characteristics, particularly a favorable roof configuration (southern exposure and no shading);
- Perception of other private benefits, such as reduced fuel bills for a number of years;
- Membership in environmental organizations and a desire to help with the energy shortage;
- Encouragement from solar manufacturers, solar dealers, friends, neighbors, and household members.

The study[2] concluded that "certain incentives will trigger household decisions to acquire solar collectors when an appropriate social network en-

---

[2] The factors that will motivate society to apply photovoltaic technology or any other solar technology, are not well understood as yet—especially as related to the early stages of commercialization. This study, which examined the psychology of early market activity, may be of interest: Seymour Warkov, *Solar Adopters and Near Adopters: A Study of the HUD Solar Hot Water Grant Program.* (Storrs: University of Connecticut, 1979).

courages that commitment and when respondents are persuaded that private, not social, benefits accrue to the household from acquiring solar," and that "critical social and behavioral experiences intervene before serious interest" results in a decision to buy.

FURTHER READING

Baldwin, Thomas E.; Hill, Lawrence G.; Santini, Danilo J.; and Stenehjem, Erik J. *Economic and Demographic Issues Related to Deployment of the Satellite Power System (SPS)*. PRC Energy Analysis Company and Argonne National Laboratory. ANL/EES-TM-23. Springfield, Va.: National Technical Information Service, 1978.

Beatty, J. Kelly. "Solar Satellites: The Trillion Dollar Question." *Science 80,* December 1980, pp. 28–33.

Bycer, Max. *Automated Solar Module Assembly Line: Quarterly Technical Report No. 5.* DOE/JPL 955287-5. Horsham, Pa.: Kulicke and Soffa Industries, Inc., 1980.

D'Alessandro, Bill. "Villagers Light the Way: Solar Cell Power in Gunsight, Arizona." *Solar Age,* May 1979, pp. 34–37.

Kortman, P. J. "Flat Plate Photovoltaic Central Station Plant, Year 2000 Characterization." McLean, Va.: TRW, Inc., 1979.

Metz, William D. "Energy Storage and Solar Power: An Exaggerated Problem." *Science* 200, June 1978, pp. 1471–1473.

Murray, William J. "Photovoltaics: Coming of Age?" *Solar Engineering,* January 1980, pp. 23–25.

"Ohio Daytimer Looks to the Sun for Its Power." *Broadcasting,* 9 July 1979, pp. 51–52.

Palz, Wolfgang. *The Prospects and Problems of Production, Storage and Terrestrial Applications of Electric Power Through Photovoltaic Solar Cells.* Prepared for the United Nations. Brussels: Commission of the European Ccmmunities, Directorate General for Research, Science and Education, 1979.

Photovoltaic Energy Systems: Program Summary. U.S. Department of Energy, Assistant Secretary for Conservation and Solar Energy. DOE/CS-0146. Washington, D.C.: U.S. Department of Energy, 1980.

Rappaport, Paul. "The History and State-of-the-Art of the Solar Cell." Princeton, N.J.: RCA Laboratories (undated). Available in mimeographed form from Photovoltaic Energy Systems Division, U.S. Department of Energy, Washington, D.C.

Riley, Julia D.; Odland Robert; and Barker, Helen. *Standards, Building Codes, and Certification Programs for Solar Technology Applications.* SERI/TR-53-095. Golden, Colo.: Solar Energy Research Institute, 1979.

Smits, Friedolf M. "History of Solar Cells." *IEEE Transactions on Electron Devices,* ED-23 (July 1976): 640–643.

Solar, Geothermal, Electric and Storage Systems Program Summary Document: FY 1980. U.S. Department of Energy, Assistant Secretary of Energy Technology. DOE/ET-0102. Washington, D.C.: Government Printing Office, 1979.

System Tests and Applications: Photovoltaic Program. U.S. Department of Energy/PRC Energy Analysis Company. HCP/T4024-01/15. Springfield, Va.: National Technical Information Service, 1979.

"The Future of Solar Electricity, 1980–2000; Developments in Photovoltaics." Report No. M101. Gaithersburg, Md.: Monhegan, Ltd., 1980.

The Final Proceedings of the Solar Power Satellite Program Review, April 22–25, 1980, Lincoln, Nebraska. DOE/NASA Conf. 800491. Springfield, Va.: National Technical Information Service, July 1980.

Warkov, Seymour. *Solar Adopters and Near-Adopters: A Study of the HUD Solar Hot Water Grant Program.* A Report to the Northeast Solar Energy Center, Cambridge, Mass. Storrs, Conn.: University of Connecticut, 1979.

Wolf, Martin. "Historic Development of Photovoltaic Power Generation." Philadelphia, Pa.: University of Pennsylvania (undated). Mineographed copies available from Photovoltaic Energy Systems Division, U.S. Department of Energy, Washington, D.C.

CHAPTER 5
Economics

The economic characteristics of photovoltaic use can be considered from the perspectives of the individual user, the photovoltaic manufacturer, the utilities manager, and the other players in the worldwide mixed economy in which solar cells are just one of many competing responses to the current energy situation. The changeover to photovoltaics and other renewable energy sources will accelerate as costs decline and as the nation's physical plant is steadily replaced. Photovoltaics will function as a technological "substitution" on the supply side of a domestic and world economy, and the magnitude of such a substitution is great: analogies are, say, the steamship for the square rigger, or the car for the horse.

The substitution of photovoltaics for conventional electrical sources will proceed over time to the limit of usefulness, when we can expect another dynamic equilibrium to be established. Subsequently, photovoltaic use is likely to fluctuate marginally in response to microeconomic considerations.

How rapidly the substitution takes place and at what level market equilibrium will occur depend on two key factors—the federal government's policy commitments to alternative energy sources and public interest in photovoltaics.

There is, in essence, no technological limit on the use of photovoltaics.

There is, in essence, one physical limit on photovoltaic use: the amount of sunlight incident on each photovoltaic array.

There is, in essence, one economic determinant of photovoltaic use: the cost of electricity produced by other means.

There have been many forecasts for the economics of photovoltaics. All conclude that before 1990 photovoltaic systems can generate electricity that is competitive with conventionally fueled power.

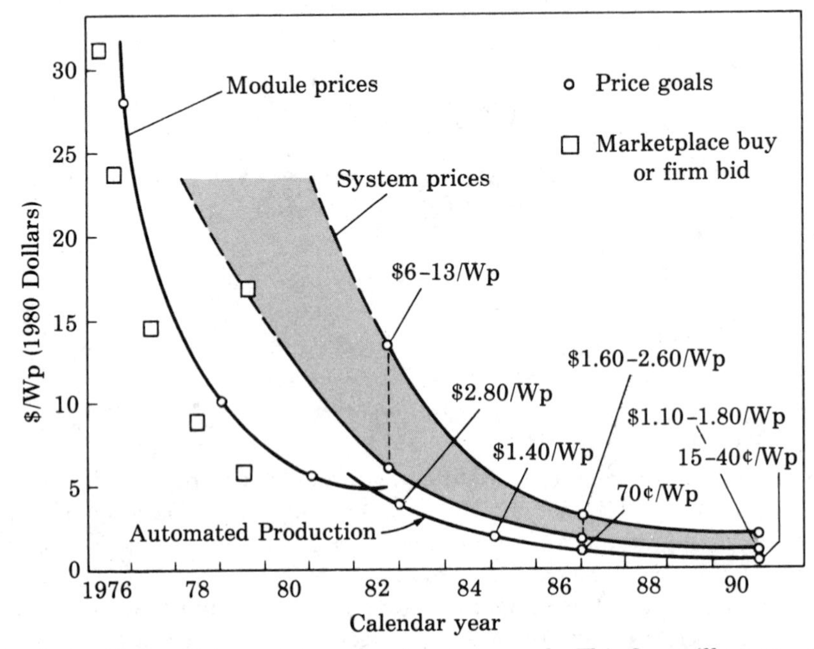

Fig. 5-1 Photovoltaic module and system price goals. This figure illustrates the drop in photovoltaic module and system prices that is expected to occur as a result of technological advances from research and development efforts, the advent of automated solar cell production lines, and to a lesser extent, the economies inherent in larger sales volumes.

Note that in the chart the price of modules is included in the system prices portrayed. The two-part module price curve is typical of price movements which occur during a technological shift—in this case, the shift from custom building to automated production of modules. Actual purchase prices of photovoltaic modules have consistently met DOE goals during the last five years.

One of the most succinct forecasts for the economics of photovoltaics is included in the multi-year program plan issued by DOE.[1] The DOE goals are specific and aimed at specific markets. The result of extensive analysis by DOE contractors, these goals are a compromise among several detailed analyses of what cost reductions in the many technical options could occur as a function of time and of resources allocated. These market analyses by DOE were based on careful surveys of the best technical judgments in industry about what is possible. Figure 5-1 depicts these relationships in graphic form. If the goals can be met, photovoltaic systems can provide

[1] U.S. Department of Energy, Assistant Secretary for Energy Technology, Division of Solar Technology, *National Photovoltaics Program: Multi-Year Program Plan* (draft) (Washington, D.C.: U.S. Department of Energy, 1980).

several hundred megawatts of generating capacity by 1986, and meet 10–15 percent of the nation's electrical needs by 2000. This is approximately the proportion of power currently generated by nuclear plants. Industry performance has consistently exceeded earlier forecasts and, if markets grow somewhat more quickly than expected, the industry may well exceed these price goals, too.

Massive deployment of photovoltaics is inevitable. As the cost of producing solar cells declines and as the nation's physical plant is replaced, more and more users will choose photovoltaics on economic grounds alone. Forecasts differ on the question of how rapidly this deployment will take place.

The market for photovoltaic devices will develop in segments rather than all at once. There is no single price at which photovoltaics becomes economic. Photovoltaics is already competitive for some uses. As photovoltaic prices drop and the costs of alternatives rise, class after class of electricity users will turn to solar devices (Fig. 5–2). This sequential, segmental market pattern is typical of newly introduced technologies.

This is the hidden story of photovoltaics: if these DOE goals are attained, homeowners can expect to install photovoltaic arrays as soon as 1986 and come out ahead financially. (See Table 1.) It should be pointed out that when we speak of economic goals and viability for residential photovoltaic systems, we intentionally exclude the effects of the federal and state tax credits available for solar purchases by individuals. In 1980, the federal income tax credit, direct tax reduction for the consumer, is 40 percent of the first $10,000 purchased.[2] In recent residential photovoltaic experiments in Arizona, another $2,000 rebate was obtained from the state income tax. This combination could reduce the installed price to the homeowner by as much as 60 percent. Photovoltaic systems that cost $2,000 per peak kilowatt generate electricity at $.06–$.10 per kilowatt hour.

The actions of the government will be the most crucial in determining how rapidly the solar transition takes place. The government's role in the political economy is unique: by virtue of its power to tax, to spend, to mandate, and to prohibit, the government is a powerful force in determining the mix of energy sources the public will use.

Renewable energy sources, with photovoltaics the most prominent, provide our only real opportunity for restructuring the energy market on a

---

[2] U.S . Congress, House of Representative, *Crude Oil Windfall Profit Tax Act of 1980,* Pub. L. 96-223. 96th Cong., 2nd sess., 1980, H.R. 3919. Section 202(a) amends 26 U.S.C. 44C(B)(2) to provide a credit against the federal personal income tax of 40 percent of the first $10,000 of residential renewable-energy-source expenditures.

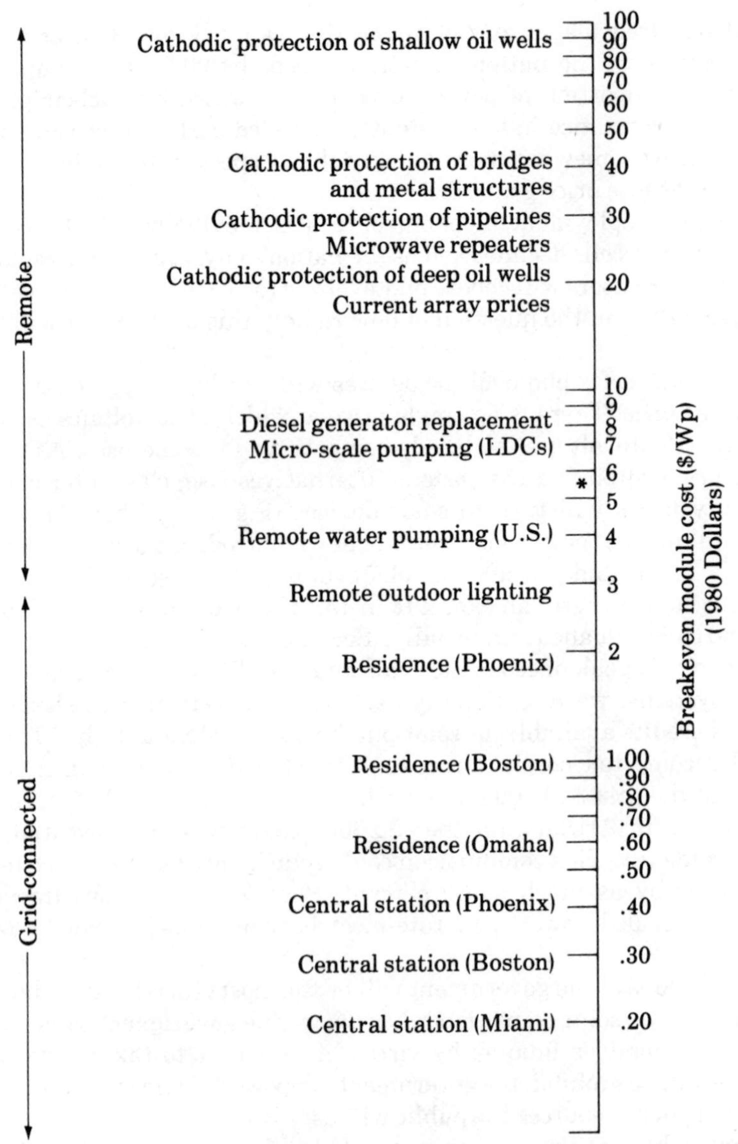

Fig. 5–2 Photovoltaic applications and breakeven costs. For each application shown, the amount indicated on the scale is the approximate price in 1980 dollars of photovoltaic modules that would make the electricity produced directly cost-competitive with electric power now used in that application. Equivalent system prices would be about double the module price. At the end of 1980, modules were selling for about $7/Wp, systems for about $14/Wp.

Table 5–1 DOE Goals in 1980 Dollars

Date	*Module Cost ($/Watt)*	*Balance of System Cost ($/Watt)*	*Total System Cost ($/Watt)*	*Net Cost of Electricity ($/kWh)*	*Major Photovoltaic Market*
Present	6.00–12.00	4.00–13.00	10.00–25.00	1.00	Remote stand-alone and micro-generation
1982	2.80	3.00–10.00	6.00–13.00	.25–.50	Diesel replacement
1986	.70	.90–1.50	1.60–2.20	.05–.09	Residential with utility backup
1990	.15–.40	.60–1.20	1.10–1.30	.04–.08	Central utility and all distributed uses.

scale large enough to lessen our dependence on all fossil fuels, and especially on imported oil.

THE INTEREST GROUPS

Before looking at the particular situations of users and suppliers, and at the overall markets, a simple summary of the economic motives of all the parties involved may help explain likely market developments.

Users. Almost everyone uses electricity, and uses it for a variety of purposes. But for whatever purpose, all classes of users are interested in acquiring needed quantities at low costs for operation, for maintenance, and for capital equipment. All users have a certain level of use below which their other activities are unacceptably hindered. Within a higher range of use, conservation measures are not burdensome, and there is a still higher level for each user beyond which consumption of electricity is clearly profligate. Users will want to get as much electricity as they can conveniently use and at as low a cost as possible.

An especially important way to divide users into subclasses is by their ability to afford the relatively large initial capital outlays which photovoltaic systems require. Many small businesses and lower- and middle-income households, in the absence of incentives, simply won't be able to allocate current income to install a photovoltaic array, even though they could realize massive savings over the life of the system. These users will benefit as the price of photovoltaic systems declines. For this to occur, others who are financially able to do so must support early market growth.

Photovoltaic manufacturers. The solar industry will want to see photovoltaic devices replace other energy sources, particularly the coal, oil, gas, and nuclear fuel used for electricity and heating. For the substitution to occur, photovoltaic manufacturers need to price their product competitively with other sources. Initially, they need to achieve great savings in the manufacturing process; the industry is just now on the verge of installing automated assembly lines.

Photovoltaic distributors. Distribution and installation of photovoltaic devices will likely be performed by small businesses whose economic motives will be conventional. Many existing heating, plumbing, and home improvement contractors may voluntarily seek the specialized, but relatively unsophisticated, training necessary to learn how to install photovoltaic systems in new and existing structures. A number of entrepreneurs specializing in solar systems will enter the field. A few nationwide firms, either new or existing, will enter the fray, using mass advertising and, eventually, their good reputations to attract sales; some will operate as franchises. Current energy suppliers, such as fuel oil distributors, may add solar sales and service capabilities in order to survive.

Local governments. States, counties, and municipalities typically wish to preserve and enhance the economic health of their areas, mainly by increasing the total spendable income of their constituencies. If solar systems can provide energy at less cost than other alternatives, we can expect to see a series of tax incentives for solar installations designed to create jobs and to free up personal income for other uses. The solar industry could provide a number of new jobs, an important aspect of economic health at the community level. State and local measures to attract producers and to encourage distribution will be enacted. But states which are large producers of coal, oil, and gas could foresee economic loss if the need for these industries is lessened. Consequently, there may be at least a minimal antipathy at the state level.

Utilities and utility regulators. The posture of the utilities is the major unknown factor in the entire photovoltaic equation. Most, if not all, of the utilities in this country are monopolies regulated by public commissions and they serve two functions which were heretofore inseparable: the production of electricity and the distribution of electric power. We see distribution as remaining a natural monopoly. Our society will still need an electric power grid and it is wasteful to have more than one grid serving any specific geographical area. But the function of production could be radically altered: every photovoltaic user could also be a generator. Whether the utilities will fight to preserve their function as the central and sole producer of electricity is open. Central generators will continue to be needed as backups during long periods of adverse weather and at times of extra-large peak demands. Utilities themselves can be expected to install centralized photovoltaic arrays for extra generating capacity whenever the cost of photovoltaic devices is less than the new cost of other kinds of systems.

Regulatory agencies are generally charged with ensuring that a utility not earn more than a fair rate of return on its operations. Whether, during a massive change in the technological basis of the industry, many or most of these agencies will interpret this change as a mandate to protect the prior investment of utility owners is unknown, but certain to be a public issue.

The federal government. The government's role, as stated, is unique. We assume that government is responsive, but is not monolithic: different parts react in different ways to different pressures, of which the economic interest of the single voter is but one. A chief motive of the government's formulation of an energy policy is to erase the potential threat to the political and economic health of the country from having large parts of its energy supply lie under the control of other sovereignties. Any strategy which lessens this reliance would be welcomed. Secondarily, the government appears to want to ensure growth of energy supplies, preferably at the least cost in development and operation; to create new jobs while not eliminating existing ones; to prevent environmental deterioration; and to enhance the physical health and safety of the public.

ECONOMICS OF PHOTOVOLTAICS ON THE DEMAND SIDE

The cost of photovoltaics, now and in projections of future development, depends on how rapidly the cost of competing fuels rises; the speed with which photovoltaic hardware is developed; and the strategies the federal government uses to foster, or to discourage by inaction, the inevitable changeover to photovoltaics. Projections depend on the definition of costs, the values assigned to external costs such as pollution or storage of radioactive waste, and such intangibles as aesthetic considerations and convenience. The price depends on the money market and on the user's tax status—the numbers look very different from the respective viewpoints of the commercial utility, the private homeowner, the renter, the landlord, the retail business owner, the manufacturer, and the non-profit institution.

Since photovoltaic systems are presently economical only for small remote applications or for specialty uses where cost is a minor consideration, what is the likely scenario for reducing the price of installed photovoltaic systems to the point that they become cost-competitive for major applications? And what figure defines that point?

For a number of substantive and technical reasons, it appears that a module price of $.60–$1.00 per peak Watt (Wp), entailing a total system price of $1.50–$2.50 per installed peak Watt, is the benchmark figure at which photovoltaics becomes competitive for general use in most parts of the United States. At this price solar arrays will generate electricity in the residential application at a full cost to the user of $.05–$.09 per kilowatt hour.

For all users, the economic attractiveness of photovoltaics will depend primarily on the combined effects of insolation rates and local electricity prices. Insolation rates determine the size of an array needed to produce a desired amount of current; thus they are a main determinant of the installed capital cost of a photovoltaic system. A house in New England would require a system about 40 percent larger than the same house in Arizona, excluding the climate-influenced differences in the power required for heating and air conditioning. See Appendix 9, Estimating Photovoltaic System Requirements for a Residence.

Local electricity prices are the key variable in operating costs. The rates at which utilities not only sell but buy back electricity appear to be critical factors in determining photovoltaics' economic viability. The decision to go solar turns on the question of whether the user will save enough money by not buying electricity to offset the price of installing photovoltaic equipment.

Economic factors such as the tax posture of the purchaser, interest rates, and government incentives play a lesser, though significant, role in determining how economically attractive photovoltaics will be. Particularly during the next decade while the technology is new and the array price is high, these other factors will comprise the determining margin. The most likely residential purchaser will be the homeowner (but not the landlord) who lives in a region with abundant sunlight, is in a high tax bracket, can borrow money at low rates, and receives a large credit for the installed price against his federal income taxes. Others less favorably situated would delay a decision to convert.

Photovoltaic systems are characterized by relatively large capital cost, low operating and maintenance cost, and free fuel. Thus, any user considering a photovoltaic system will have to compare the cost of capital, amortized over the life of the system, against the projected expenditures for purchasing electricity from a central utility. In the absence of compensatory incentives, this high-capital requirement militates against those who cannot afford to wait to recoup their investment and those, such as renters, who have no capital invested in the site and do not expect to stay long enough to realize any savings on operating costs.

When arrays become available at \$.70 per peak Watt of generating capacity, the DOE goal for 1986, installed photovoltaic systems will cost \$1.60–\$2.20 per Watt and provide electricity at a total system-lifetime cost of \$.05–\$.09 per kilowatt hour. This price is completely competitive with the 1979 average residential cost of centrally generated electricity in the United States, about \$.05 per kilowatt hour.

Each photovoltaic application will have a unique economic configuration. A look at the numbers involved in two uses of photovoltaic power, replacing a diesel-fueled generator and supplying a U.S. residence, will indicate the validity of this breakeven price range and illustrate the calculations involved in deciding to convert to photovoltaics from a total reliance on the grid.

The Small Portable Generator

Remote stand-alone or portable applications are now fully economic when used to recharge, or extend the life of, battery systems. About 3 megawatts of photovoltaic systems are sold each year for this use and the world market could grow to as much as 500 megawatts per year.[3] At present, battery systems are frequently recharged by small diesel or gasoline generators. These generators ordinarily produce less than 50 kW of electricity and are used for such purposes as powering small remote villages, small radar installations, and mobile Army units.

Breakeven cost analyses for a diesel generator compared with a photovoltaic system have been developed in detail by the NASA Lewis Research Center[4] and by Royal Dutch Shell in the Netherlands.[5] Both studies concluded that photovoltaic systems are competitive with small diesel generators when the installed photovoltaic cost is less than \$5 per peak Watt.

At present, installed costs for a photovoltaic direct current, no-storage, replacement for a diesel generator (with the diesel as a backup) is about \$10 per Watt. This means that while there must be a factor-of-ten reduction in cost for photovoltaics to be directly competitive as a general power source in most parts of the U.S., solar cells are now within a factor of two from full economic viability for this particular application.

---

[3] Private communication, Royal Dutch Shell.

[4] Louis Rosenblum, William J. Bifano, William A. Poley, and Larry R. Scudder, *Photovoltaic Village Power Application: Assessment of the Near-Term Market,* prepared for U.S. Department of Energy, Division of Solar Energy. DOE/NASA/1022-78/25, NASA TM-73893 (Cleveland: U.S. National Aeronautics and Space Administration Lewis Research Center, January 1978).

[5] Private communication, Royal Dutch Shell. Proprietary study.

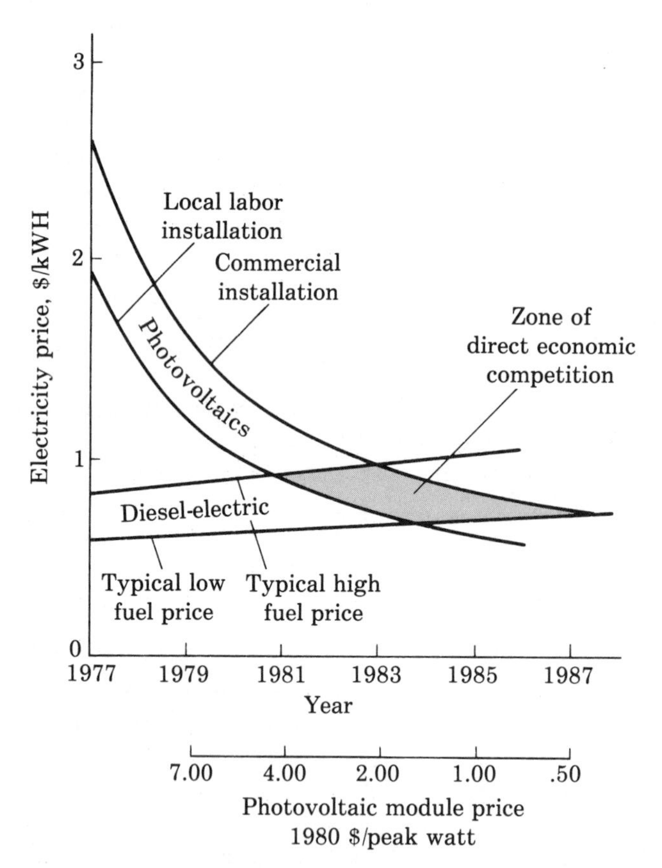

Fig. 5–3 Energy price comparisons for diesel generator and photovoltaic 1-kW continuous power systems. This figure compares the energy prices of a 3-kW diesel generator and a 5.5-kWp solar photovoltaic system with battery storage. Both produce 1-kW continuous power, and the photovoltaic system requires relatively good insolation. The two diesel curves reflect fuel prices of $.60 and $1.50 per gallon, rising 5 percent per year. The two photovoltaic curves reflect the use of local noncommercial and commercial labor, costing 10 percent and 50 percent, respectively, of the total installed system cost. Both photovoltaic curves reflect the module prices shown on the lower scale. The overlapping sections of the curves denote the economic breakeven zone. Diesel fuel prices have been rising faster than the projections shown here, and photovoltaic systems are now producing electricity competitively with diesel generators in many parts of the world.

The cost of electricity from the diesel generator can be compared to the cost of electricity from a photovoltaic system. The principal cost of operating a diesel generator is the fuel it consumes. To produce 2,000 kilowatt hours of electricity, a well-maintained 3-kilowatt diesel generator (producing 1 kilowatt of continuous power) burns about 400 gallons of fuel.[6] Figure 5–3 shows the effect of fuel cost on electricity generated in this way.

According to Royal Dutch Shell, the average price for delivered diesel fuel in remote parts of the world's Sunbelt is $2–$3 per gallon. This means that the 10 million small diesels in use throughout this area are generating electricity at a fuel cost of $.40–$.60 per kilowatt hour. If we amortize the cost of the generator and add in the cost of spare parts and regular maintenance, the cost of this electricity is higher, typically ranging from $.50 to $1.00 per kilowatt hour (Fig. 5–4).

If the same amount of electricity were generated by a photovoltaic system costing $10 per installed peak Watt, a complex new set of economic factors must be considered. Since a capital asset requiring no fuel would be used to generate electricity, to portray accurately the cost of photovoltaic power we need to know the cost of money, the nature of financing, the amount of sunshine falling on the array, and whether storage is required. To understand the effects of these variables on costs, the cost of photovoltaic electricity from the $10,000 per peak kilowatt assumed earlier will be developed as a function of different values.

The assumptions are:

- A Sunbelt site where the annual solar flux is sufficient for the photovoltaic generator to produce 2,000 kilowatt hours per year per kilowatt of generating capacity;
- A 20-year loan with a 15 percent interest rate;
- A photovoltaic system life of 20 years;
- Maintenance costs of 1 percent of capital per year;
- Peak capacity of 1 kilowatt (photovoltaic systems have little or no economy of scale).

If the installed capital cost is $10,000 (1980 dollars) and this one-kilowatt generator produces 2,000 kilowatt hours per year, to find the cost of electricity generated the annual cost of capital plus maintenance is divided by the system's output in kilowatt hours per year.

The annual capital cost to amortize the mortgage and maintain the system is nearly 20 percent of $10,000, i.e., $2,000. Thus, the cost of electricity per kilowatt hour (kWh) is $1.00. This compares favorably with

[6] Louis Rosenblum et al., *Photovoltaic Village Power Application: Assessment of the Near-Term Market*, p. 21.

60-kilowatt photovoltaic system
Lifetime: 20 years—131,400 kWh/yr
2,366 modules (97,000 cells)
½-acre array
20-year cost (1980 dollars):

Capital investment (land additional)	$600,000
Fuel	0
Total	$600,000

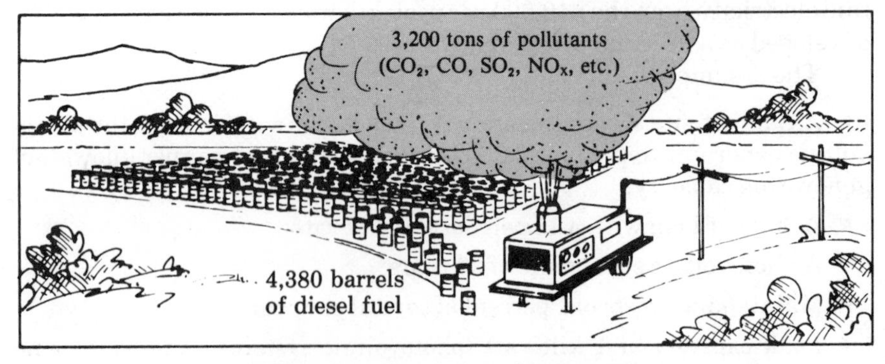

60-kilowatt diesel-electric generator
Lifetime: 20 years—131,400 kWh/yr
Diesel replacements: 10 @ $16,000 each
Fuel: 9,000 bbl (1,039 tons)
20-year cost (1980 dollars):

Capital investment	$160,000
Fuel (@ $3/gal)	$1,134,000
Total	$1,294,000

(operation and maintenance extra)

Fig. 5–4 Comparison of photovoltaics with diesel-electric for power generation in remote areas. Because of the high price of diesel fuel in developing countries, photovoltaic systems are fully competitive with diesel-electric generation in 1981 for many areas. (Idea courtesy of Andrew Krantz)

the cost for fuel as given in Figure 5–3. If a cost of $5,000 for the photovoltaic system had been assumed, the cost per kilowatt hour would be about $.50, and therefore fully competitive with the diesel generator. This halving of costs is not only probable, but virtually certain.

The Private Residence

Photovoltaics is well on its way to supplementing electricity now generated by central fossil- or nuclear-fueled utility plants. Several hundred residential photovoltaic systems will be purchased for private use in 1981, and by the end of the century, millions of residences will be served by individual, semi-independent photovoltaics systems. (To some extent this discussion can be generalized to commercial and industrial applications as well, since the expense of the photovoltaic plant can be amortized and deducted as a legitimate expense of doing business.) How is this photovoltaic penetration into the market going to come about?

Going back to the DOE goal for 1986, photovoltaics will be competitive in many areas of the market when modules cost about $.70 per Watt of generating capacity, with installed system cost of $1.60–$2.20 per peak Watt. This translates into photovoltaic electricity at a cost of $.05–$.09 per kilowatt hour. But photovoltaic systems capable of providing 60–80 percent of the electricity required for a typical new residence can be fully economic even before this benchmark is reached. For instance, when total installed system costs are $1,500–$2,000 per peak kilowatt ($1.00–$2.00 per peak Watt), and there is insolation equivalent to that in Phoenix, and electricity from the utility grid costs at least $.06 per kilowatt hour (1980 dollars), photovoltaics will be cost-competitive. With New England insolation, photovoltaics will be cost-competitive when electricity costs $.08 per kilowatt hour.

In a world of rapidly changing fuel prices, housing costs, mortgage rates, and incomes, it is nearly impossible to describe a typical residential situation. Still, in order to understand how these variables interact in determining when photovoltaics becomes economically feasible, discussing a representative case can be useful. For purposes of discussion, break-even is defined as the point at which the monthly cost of owning and maintaining a photovoltaic system equals the value of the electricity that would have been purchased from the grid or generated by some alternative system had the photovoltaic system not been installed. This point is determined by comparison with utility rates at a given location. Assume that:

- You are about to buy a 1,500-square-foot, three bedroom, single family, four person, one story house with a south-facing roof unshaded by trees.

The house is well insulated and properly designed for passive solar heating and cooling.

- The house costs $65,000 without the photovoltaic system, and as buyer, you qualify for a 20 percent down, thirty year, 12 percent mortgage.
- Your income is $25,000 per year, a level at which the federal tax on your income, and the tax rebate for mortgage interest is 35 percent.
- Your use of electricity averages 1.5 kilowatts per hour, which equals 36 kilowatt hours per day, 1,080 kilowatt hours per month, and 13,140 kilowatt hours per year. The house is not heated by electricity and there are no storage devices such as batteries. If the cost of electricity is $.06 per kilowatt hour, your electrical bill is $65.70 per month, or $788.40 per year.
- The monthly payment on the thirty year, 12 percent, $52,000 mortgage is $10 per $1,000 per month, or $520.00 per month.

Now add in the cost of the photovoltaic array and support system, making these assumptions:

- Size of System: The system can be sized to provide any percentage of the annual electrical usage. A system providing 70 percent of the electricity consumed annually is desired. If the utility serving the area has a cooperative attitude toward renewable energy systems, all excess photovoltaic power you generate is sold to the grid, and all backup power is provided at average residential price to the photovoltaic system-powered residence. In Phoenix, such a system will generate 2,500 kilowatt hours per year for each peak kilowatt of installed capacity. Since 70 percent of the annual usage of 13,140 kilowatt hours is about 9,200 kilowatt hours, and one kilowatt peak capacity of photovoltaic system can produce 2,500 kilowatt hours, the system size for 70 percent of annual usage is 9,200/2,500 or 3.68 kilowatts.
- Installed System Cost: The federal photovoltaic program envisions installed residential photovoltaic systems costing $1,600–$2,000 per peak kilowatt. Therefore the installed cost of the 3.68-kilowatt system ranges from $5,900 to $7,400. For this discussion, use $7,000 as the cost of the installed photovoltaic system.
- Monthly Cost of the Residential Photovoltaic System: The 40 percent tax credit is ignored for the moment. If you qualify for a loan of $59,000 rather than the $52,000 first proposed, the additional monthly cost for the photovoltaic system would be $70, and the annual cost $840. In the first year of a thirty year mortgage, virtually all of the cost is tax deductible. The homeowner's income tax deduction, therefore, is 35 percent of $840, or $294. This yields a first-year cost for the photovoltaic system, which provides 70 percent of the electricity, of $546 per year or $45.50 for 767

Table 5–2 Key Economic Factors in the Purchase of Photovoltaic Systems Compared with Grid Electricity in the Southwest U.S.

Factor	*Residence*	*Industry*	*Central Utility*	*Municipal Utility*	*College or University*
Cost of grid electricity	5–9¢/kWh	2–5¢/kWh	3–4¢/kWh	3–4¢/kWh	6–8¢/kWh
Cost of money* (Prime less 3%)	9% Tax Rebate in addition	14% (Prime + 2%)	15% (Prime + 3%)	3–5%	5–7%
Term of loan (yrs)	30	20	20	40	40
Land or roof cost	Free (no extra cost)	$1/sq. ft.	$1/sq. ft.	Free to some	$1/sq. ft.

* A prime rate of 12% is assumed for purpose of illustration.

kilowatt hours per month, or about $.06 per kWh. (Obviously in the latter years of the mortgage the amount of deductible interest decreases.)

Such a representative case has this result: photovoltaic cost per month for the first year is $45.50. The same amount of electricity when purchased from the grid at $.06 per kilowatt hour costs $46.00; at $.05 per kilowatt hour, the cost is $38.33. Given the assumptions made, the cost of photovoltaics compared to grid electricity at $.06 per kilowatt hour is breakeven. The numbers would be different had the house been purchased in New England, but the results are much the same. The size of the photovoltaic system would need to be larger (about 6 kilowatts peak capacity) and it would cost $10,000–$12,000. Monthly costs of the mortgage to finance the array and other components would be as low as $100 less the $35 federal income tax deduction, or $65 per month for 767 kilowatt hours per month. The breakeven for a New England home is at $.085 per kilowatt hour grid electricity.

In 1980, installed residential photovoltaic systems cost $10,000–$15,000 per kilowatt hour peak capacity. The 40 percent federal tax credit enacted in 1980 for purchasers of photovoltaic systems will enable more locations to become fully competitive provided the utility company cooperates by providing backup electricity at a fair price.

The cost of photovoltaic systems, the cost of money, tax status, and cost of grid electricity are all parameters that must be factored into the decision as to when purchasing photovoltaics is economically advantageous. Table 5–2 shows five different applications in the Southwest, and compares the key costs involved in the purchaser's decision. The homeowner is in a more favorable economic position to buy a photovoltaic system than other users of electricity, or than electric utilities who produce it. He or she can borrow money at lower rates, deduct the interest paid on a loan in computing personal income tax, and in addition is allowed to subtract directly from the computed income tax 40 percent of the cost of the photovoltaic system. At the same time, the homeowner pays more for electricity that he buys. For this reason, residential applications may represent the first major domestic market for photovoltaics.

If you build or buy a new home, several key cost factors favor an early decision to purchase photovoltaics. For instance, your thirty year home loan has a low interest rate compared to the rates paid by industry, and when you compute your personal income tax, the interest is deductible. Depending on income, this effectively decreases the cost of the borrowed money to as low as 6 to 8 percent (assuming a loan rate of 12 percent). In contrast, the industrial customer or central utility must pay 2–3 percent more than prime rate for new capital—i.e., 14–15 percent. This is a relative near-term disincentive for the business or utility concerned. On the other hand, a university or other school can obtain construction funds for as little as 3 percent interest, and spread repayment of the loan over forty years. On this variable alone—the cost of money—there is a spread of a factor of five for identical solar-electric hardware installations.

Another key factor is the cost of the electricity replaced. From the point of view of a central utility manager, the value of electricity with which photovoltaic costs should be compared is his marginal cost of generating electricity from a new plant that uses some alternative source: coal, nuclear, hydro, whatever. That is, the comparison is in terms of the cost of generating more electricity from a new plant. This now ranges from $.03–$.08 per kilowatt hour. But the homeowner or buyer who is considering a residential photovoltaic installation must make decisions based on the price of the electricity he buys from the grid. This ranges from $.02–$.14 per kilowatt hour in the U.S., roughly twice the price per kilowatt hour in the utility manager's situation. On this basis the homeowner or buyer, then, will find it profitable to install a photovoltaic system much sooner than the utility manager.

Still another factor in the decision process is whether the cost of electricity is an expense item in the tax structure of the customer. The manager in business or industry can deduct all electricity costs from his income; in effect, he is receiving a 45 percent rebate from the IRS for using non-renewable fuel. If he were to purchase capital to create electricity from the sun, however, IRS may allow as little as 5 percent of capital per year to be depreciated.

The solar investment tax credit mitigates this factor somewhat. Tax-exempt institutions, because of their status, can deduct neither power costs nor capital. The homeowner can deduct all interest on the home loan to purchase the capital for the photovoltaic system, but cannot deduct the cost of purchasing electricity from the utility.

Obviously, identical photovoltaic systems can have very different costs and benefits, depending on the purchaser's status vis-à-vis a number of variables. Still, it appears that when everything is factored in, the individual home will be the first major market for photovoltaic electric systems in the United States.

In the representative case discussed, conservative assumptions were used: that money will continue to be costly to borrow; that oil prices (and prices of other fuels that "float" according to oil's dearness) will not continue to rise at the rate of the past few years; that there will be no breakthrough technology, i.e., improvements in photovoltaic systems are assumed to be incremental; and that photovoltaic systems will last twenty years, although there is no theoretical reason why they should not have longer useful life, given reasonable care. With less restrictive assumptions, photovoltaics is cost-competitive for some homeowners today—if the homeowner is in the 50 percent tax bracket, pays $.12 per kilowatt hour, receives 2,500 kilowatt hours of insolation annually, obtains a 50 percent tax rebate, and can depend on the utility grid to back up the system.

There is no single figure at which photovoltaics becomes cost-competitive. There are a range of applications to which photovoltaics can be put, each of which must be analyzed separately. Within each class of applications, the tax structure, utility costs, and income of the owner all collaborate to determine breakeven. The development of the industry itself needs to be examined within the context of a global mixed economy; there are more variables, this time on a macroeconomic level. Photovoltaic viability on a scale large enough to seriously relieve our dependence on non-renewable fuels is intimately tied to the size of its market.

ECONOMICS OF PHOTOVOLTAICS ON THE SUPPLY SIDE

A market is made up of both buyers and sellers. Multiple economic considerations affect both manufacturers of terrestrial photovoltaic equipment and utilities in their role as suppliers of electricity.

By 1984, homeowners will want to install photovoltaic systems in order to save money on their electric bills. By 1986, power companies may opt for photovoltaics to save money on generating costs.

In the 1960s, when solar cells were used mostly on spacecraft, five companies entered the field: Heliotek; Hoffman (later Centrallab, now Applied Solar Energy); International Rectifier; RCA; and Texas Instruments. The annual market grew from 7,000 Watts in 1960 to a steady level of 80,000 Watts in 1968–1970, selling at between $100 and $200 per peak

Watt.[7] Along the way, the latter three companies dropped out of photovoltaics production. A simple case of industrial overcapacity had developed: there were too many producers and not enough places, at current prices, to sell.

In the early 1970s, an awareness of a potential terrestrial-use market spread through some segments of industry, the public, and the government. By 1979, the U.S. photovoltaic industry comprised more than 100 companies pursuing five main activities: raw materials production (typically upgrading raw silicon to semiconductor-grade crystalline form); photovoltaic components production (typically producing silicon wafers from crystalline ingots); photovoltaic systems manufacturers (assembling photovoltaic modules and arrays); original equipment manufacturing (installing electric power sources on such products as wristwatches and portable refrigerators); and research and development in all the above phases.

The process begins with raw silicon. Dow Corning is the largest domestic producer of semiconductor-grade silicon. Its 500-metric-ton production in 1979 comprised about half of the output for the United States and a fourth of world total. Other producers include Texas Instruments, Monsanto, Motorola, and Great Western. About 1 percent of this silicon is used by photovoltaic manufacturers, and the cost of the silicon accounts for at least half the cost (not price) of the finished cell.[8] This manufacturing process is energy- and capital-intensive, and new production capacity using the standard Siemens process requires a long lead time, typically 3–5 years) to bring on line. Although prices are expected to fall from $60 to $10 per kilogram by 1986, many observers predict a severe market shortage of silicon in the interim, in part because potential producers are unlikely to enter the field until the imminent breakthroughs in production technology are proven.

More than twenty U.S. firms supply photovoltaic components. Mainly, these companies transform silicon ingots into wafers. The process is relatively simple and takes little time or capital to accomplish.

Some fifteen U.S. firms manufacture silicon photovoltaic modules systems for terrestrial use; two others limit their sales to space applications.

Most of these companies carry module production forward into the final assembly and sale of photovoltaic arrays; they also sell modules to other companies which specialize in final assembly. 1980 module prices

---

[7] See Martin Wolf, "Historic Development of Photovoltaic Power Generation," photocopy (author affiliated with the University of Pennsylvania, Philadelphia, Pa. [ca. 1975]), Fig. 11; and U.S. Department of Energy, Assistant Secretary for Conservation and Solar Energy, *Federal Policies to Promote the Widespread Utilization of Photovoltaic Systems,* DOE/CS-0114/2 (Washington, D.C.: Department of Energy, 1980).

[8] Press release, Solarex Corporation, 23 June 1980, Rockville, Md.

Table 5–3 Relative Share of Total Market Held by Active Producers of Photovoltaic Cells and Arrays: 1978

	Sales in Dollars (%)	*Capacity in Kilowatts* (%)
Solarex	42	45
Solar Power	16	17
ARCO Solar	10	12
Motorola	7	7
Sensor Technology	6	6
Phillips-RTC	8	5
Sharp	2	1
Other	8	7

ranged from $6–$12 per peak Watt and installed systems from $10–$40 per peak Watt. Total sales comprising nearly 4 megawatts amounted to $50 million. Table 5–3 shows the relative market shares held by active producers in 1978.

Another segment of the industry, original equipment manufacturers, is concerned mainly with the reliability and cost effectiveness of its products and electric power source. Some products now use photovoltaics, and more will do so when photovoltaics meets these criteria.

Finally, some 25 domestic photovoltaic companies are engaged solely in research and development activities. Most of these rely exclusively on federal funds. Many have a weak commitment: if funding were ended, they would drop photovoltaics work.

Several unresolved energy issues make it difficult to forecast the characteristics of a mature photovoltaic industry. One of these concerns ownership and competition. Of the fifteen or so silicon module manufacturers, several are independent, several are subsidiaries of electronics companies and two are oil company subsidiaries. ARCO Solar, formed through acquisition by Atlantic Richfield, and Solar Power, owned by Exxon, are actively selling photovoltaic arrays. Other oil companies, including Chevron and Gulf, have launched R&D efforts. Shell Oil owns SES Inc., a firm specializing in cadmium sulfide arrays. Totale Oil in France is a large owner of Photon Power, another cadmium sulfide entrant. Several other photovoltaic companies are owned or dominated by large companies. Spectrolab is owned by Hughes and Photowatt is owned by GGE in France.

There is no question that Big Oil is becoming Big Energy. The energy industry today means oil, gas, coal and nuclear. In 1979 it accounted for 20 percent of the gross national product. The oil giants now own and produce most of the coal and a significant fraction of gas and nuclear electric power, and they have also diversified into other energy and non-energy areas.

And in the last five years, Big Oil, now Big Energy, has been quietly buying up small solar energy independents.

For several reasons, expansion into photovoltaics is a logical step for the oil giants to take:

1. Over the long run, oil supplies will dry up, but the need for energy will not.
2. Like oil exploration, photovoltaics is a long-term investment, and presents no peculiar management problems.
3. Oil companies are among the largest in the country. Large profits force them to consider alternative investments, and enable them to sustain money-losing ventures for a long time. Photovoltaics presents sound opportunities for investment in the energy field which they have the capacity to underwrite.

In the photovoltaic industry, some ten companies that were independent in 1975 now have been reduced to one. Although other small companies have formed, the trend toward concentration in the energy industry is clear. Is this trend healthy, from the standpoint of providing solar photovoltaic equipment to the marketplace at reasonable prices and in the quantities that the American people want? More to the point, will the new owners, who obviously have other interests, give solar a fair boost into the marketplace? At this stage, there is no way to be sure. Much will depend on government attitudes, but even more may depend on the public's professed desires.

However, the presence of both electronics subsidiaries and independents in the photovoltaic module industry indicates a diversity which suggests that both technological and price competition will continue unabated over at least the short and medium run.

Furthermore, the cost of entering the field is low. At present, economies of scale in flat-plate array assembly level off at plant sizes of between 30 and 50 peak megawatt annual production. To start a plant of this size will require an investment of only $10 to $15 million in 1986.[9]

Since entry cost is low, however, the independent companies in the field are always vulnerable to takeover. The independents include Solarex, Solec International, and Solenergy. A fourth, Applied Solar Energy, has a large amount of public stock outstanding. Solarex, the largest of all domestic module producers, is closely held by its founders. However, in 1980 substantial minority shares were sold to Standard Oil of Indiana (Amoco).

---

[9] *Federal Policies to Promote the Widespread Utilization,* Vol. 2 (preliminary), pp. 6–8.

A second issue concerns the economic impacts of the rapidly changing technology in the field. Only two module production methods are approaching anything near maturity: single-crystal silicon and cadmium sulfide. Whether these rapidly evolving technologies or some of the others now in earlier states of investigation will become dominant in the next few decades is not now known.

This rapid pace of change yields several uncertainties. Potential investors in a new enterprise may be unwilling to gamble on an operation whose technology may soon be obsolete. Also, no one knows what the most efficient scales of operation for both present and future technologies will be, and this scale is the single most important determinant of the future structure of the industry. One study speculated, "It is possible that silicon materials production could become most efficient at large scales, module production at medium scale, and residential systems sales, installation and service at a scale attractive to small business."[10]

In any event, the industry is young enough and the technology changing so rapidly, that the economic advantages gained by small firms' proprietary R&D findings could offset the advantages held by large firms' capital resources.

A third issue is a semiconductor-grade silicon shortage expected to be felt by 1982 or 1983. This supply problem stems from the growing demand for the product by both the booming semiconductor and photovoltaic industries and the long three-to-five year start-up time needed to bring new production facilities on line. Although the market will be able to respond by the mid-1980s, the specific effects in the near term, other than higher silicon prices, are unknown. By late 1980, it appeared that the problem would not be serious.

A fourth issue concerns the nature of vertical integration of firms in the whole photovoltaic production process. Manufacturers of metallurgical silicon may move to produce crystalline silicon and vice versa. Crystalline silicon manufacturers may move forward into wafer, module, and system production. Original equipment manufacturers may move all the way back to module production. These moves will be made for financial profit and, in some cases, to ensure supplies of manufacturing components. Again, economies of scale will dictate industry structure, and it remains to be seen where these economies will lie.

A final issue concerns government policy. The government has not yet made some of the decisions that, by virtue of its role, it will have to make. These unmade decisions will also affect industry structure. The imminent

---

[10] *Federal Policies to Promote Widespread Utilization,* Vol. 2, pp. 12–48.

economic viability of photovoltaics has not even become common knowledge among parts of the Department of Energy, much less among other departments involved in energy issues, or Congress generally.

DISTRIBUTION OF PHOTOVOLTAIC DEVICES

Distribution is the final step in getting photovoltaic devices out of the research journals and onto rooftops. Distributors have the task of locating potential purchasers and installing devices on unique sites. The problems are very similar to those faced by the heating and cooling industry, and we can expect photovoltaic distribution to resemble that model closely, if not to become an actual adjunct.

A mix of the following seems likely:

Photovoltaic systems manufacturers may wholly own or may franchise a network of distributors;

Independent distributors may stock a variety of brands;

Existing heating, air conditioning, plumbing, electrical, and home repair and improvement firms may add photovoltaic capabilities;

A number of photovoltaic specialists could appear;

Large national retail firms may market photovoltaic systems and provide installation and maintenance services.

The construction industry could serve as a natural conduit for bringing photovoltaics into widespread use. However, the entire industry, including private contractors, financers, and public regulatory bodies, is notorious for its institutionalized lack of innovation. Photovoltaic deployment will be slowed until these parties become better acquainted with the opportunities it offers.

There will be barriers to photovoltaic deployment within the construction industry because:

- Housing builders are typically very small businesses, and have a strong tendency to keep doing only what they know how to do;
- There are 30,000 building code variations spread among the nation's 4,000 code-making authorities, and these codes tend to be oriented to specification rather than performance: one must "do it this way with these materials" rather than "achieve this result."
- Planning and zoning boards and building code inspectors, one by one, will face demands for photovoltaics. Many will resist.
- Bankers don't like to lend money for unknowns. News of proven photovoltaic performance will seep slowly through the financial community.

These barriers are institutional rather than technical or economic. Under pressure from consumers, manufacturers, and national energy needs, such obstacles will fall—slowly and grudgingly, perhaps, but they will fall.

UTILITIES

The utilities' physical and economic positions are unique in any energy analysis. The utilities act as both producers and distributors of electric power, and serve all areas except where the cost of erecting new transmission lines is prohibitive (new users may bear this cost if they wish).

Because it would be a massive waste to permit the erection of more than one set of central generators and transmission lines to serve a single area, governments have granted utilities the right to operate as monopolies. To protect the economic interests of consumers, governments have also created public regulatory bodies with the power to establish sales prices and the responsibility to supervise operations. Utilities are required to provide power to anyone in their area who wants to buy it, which explains their need always to have ample generating capacity. Similarly, utilities are eager to urge conservation if a power shortage is foreseen. Investor-owned utilities are allowed to make a reasonable return on their investment.

Figure 5–5 depicts the distribution of power generation by ownership. Recent production and projected growth in demand are shown in Table 5–4. There is serious doubt that electrical demand beyond 1985 could be met entirely by coal and nuclear facilities. The required plants probably could not be constructed in time, their merits aside. Diligent conservation measures would make the shortfall less painful. But photovoltaics now virtually assures that adequate power will be available from clean, safe solar energy.

In countries which already have a well-developed electrical grid, photovoltaics will not alter either the need for, or the basic expense of maintaining and operating the distribution network. We can assume that because of its convenience, the transmission grid will continue to be maintained, that consumers will bear the costs, and that public agencies will continue to regulate.

Extensive use of distributed photovoltaic systems will, however, alter the nature of electric power generation. As in the case of the potential residential user, many variables enter into the economics of central power generation. While precise analysis is difficult, the major factors are clear enough to let us say with confidence that photovoltaics could benefit both the user and the utility.

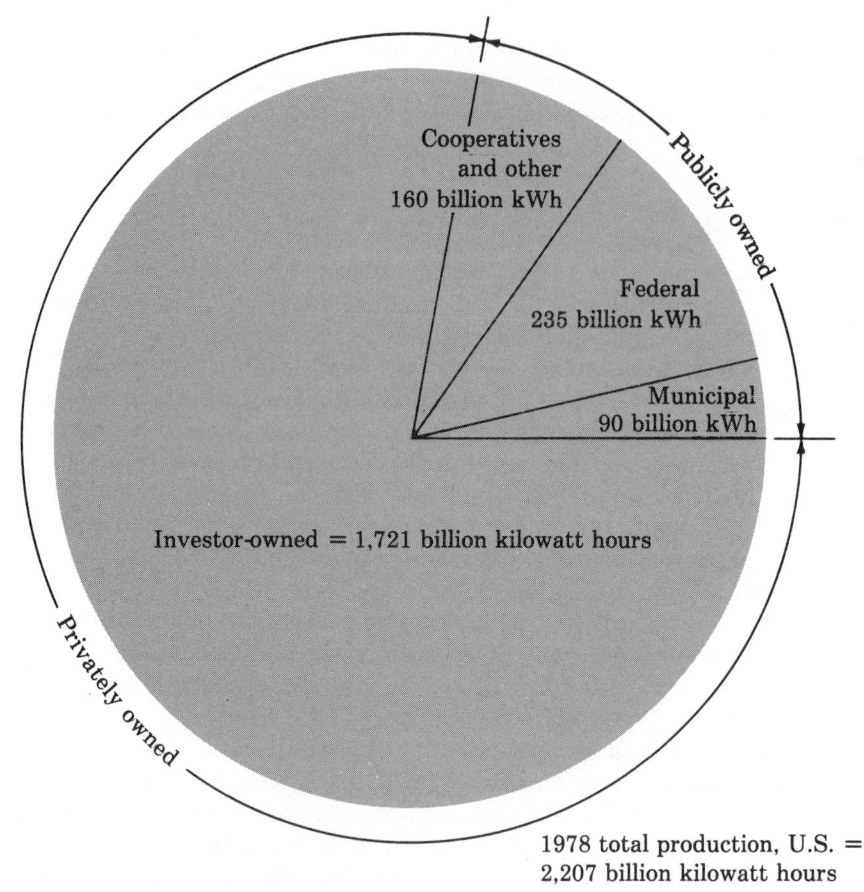

Fig. 5–5 1978 utility-produced electricity for public use, by ownership.

The two major costs of central power generation stem from plant capacity (the cost of constructing and maintaining generating facilities) and from fuel (the main cost of operating them). Photovoltaics will affect both, but by differing degrees. Such conditions as time of day, season, latitude, cloud cover, and available sunlight, and total amount of photovoltaic electricity production affect the economics of photovoltaics.

Utilities have three different levels of power production which they are required to execute: base load (the minimum amounts required); intermediate (a "normal" level of consumption); and peak (extra-heavy demands during very cold winter days when heaters are on, or very hot summer days when air conditioners are used most).

Table 5–4 U.S. Utilities' Electricity Production for Public Use

Year	*Total Production (Billions of kWh)*	*Change from Previous Year Listed (%)*	*Per Capita Production (kWh)*	*Change from Previous Year Listed (%)*
1960	753	—	4,200	—
1973	1,860	147.0	8,900	111.9
1974	1,867	0	8,800	(1.1)
1975	1,918	2.7	9,000	2.3
1976	2,038	6.3	9,500	5.6
1977	2,124	4.2	9,800	3.1
1978*	2,207	3.9	10,100	3.1
1980**	2,505	13.5	11,300	11.9
1985**	3,238	29.3	13,900	23.0
1990**	4,142	27.9	17,000	22.3

*Preliminary

**Projected using U.S. Census Bureau Series II (mid-level) population projections.

Source: *Statistical Abstract of the United States: 1979,* Tables 1025 and 1026; *Statistical Abstract of the United States: 1978,* Table 1015.

All three levels have different economic characteristics. Most notable are the high fuel costs and the use of less efficient, expensive generators associated with peak period production.

Studies both by the utility industry and by the government indicate that the overall economic effects of distributed photovoltaic systems on utilities will be very different at low (less than 10–20 percent) and at higher photovoltaic penetrations,[11] with some impacts reversing completely for example, the use of distributed storage systems.

Low penetration by photovoltaic systems will have little effect on utility production costs. As more of these early photovoltaic systems come on line, they will displace some of the generating capacity needed during base load, intermediate, and peak periods, allowing utilities to defer the construction of new generators. Photovoltaic use decreases coal, oil, gas, and nuclear fuel consumption during all three load levels, replacing conventional generating capacity by somewhere between zero and 50 percent of the rated capacity of the photovoltaic system (Fig. 5–6). Photovoltaic systems have their highest output during midday, the time when utilities experience peak loads. To meet this peak, utilities presently must bring on line their most costly generators and then return to cheaper sources when the peak has passed. These peak loads are characterized by proportionately higher fuel and lower capital costs compared to off-peak loads.[12] By

[11] *Federal Policies to Promote Widespread Utilization,* Vol. 2, pp. 2-10–2-11.

[12] Because peaking is provided by smaller, older and less efficient units, capital costs assigned are lower as the older facilities have been written off.

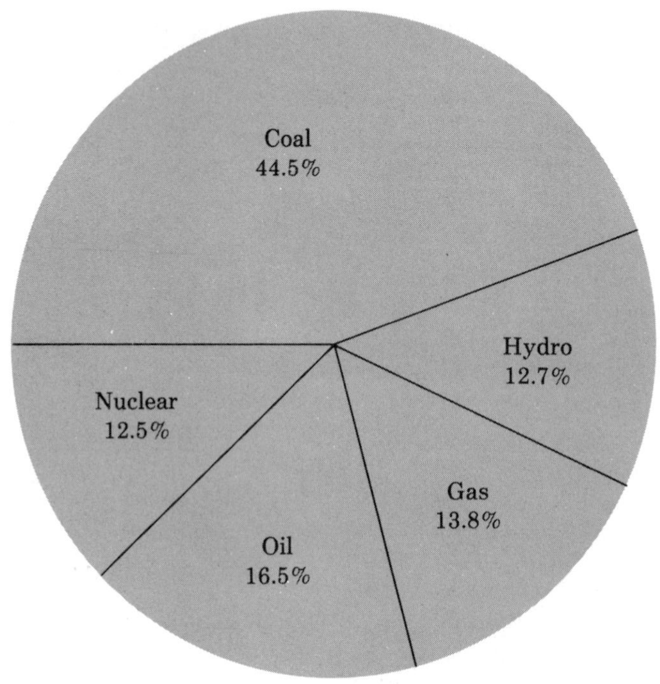

Fig. 5–6 Fuel used by utilities to generate electricity in 1978 (percentage of kWh produced).

deferring installation of new generators, the existence of photovoltaic systems could save the utilities money.

At higher penetrations (i.e., greater than about 20 percent of a utility system's total output), photovoltaics will begin to replace base load capacity. In addition, utilities would have to maintain the operational flexibility to service peak loads which would be proportionately larger than today's.

Photovoltaic systems do not require storage devices to achieve economic viability. But when the issue of storage comes into play, a whole new set of economic factors obtains.

If storage systems are distributed, the residential owner may use any excess photovoltaic-derived power either to recharge the home storage system or to sell back to the utility company. The owner can be expected to recharge his storage system first and sell back later. The utility company still receives power during peak periods and thus levels off the associated high operating costs.

If the utility itself maintains a storage system whose capacity it would discharge during peaks, then distributed photovoltaic storage and central storage economically compete.

At high penetration levels, photovoltaics alone could supply the excess needed during peaks and proceed further to recharge the utility's storage system. The utility's peak load will then occur at some other time of day, likely at the slightly lower levels around the current daytime peak, and could be served by system storage. Photovoltaic generation and system storage, then, become economically complementary. As indicated in Figure 5–7, even limited storage capability can be important.

Further, photovoltaic systems need not be distributed at all. Until recently it was thought that not before photovoltaic system prices fell to $1.10–$1.30 per peak Watt, which is expected to occur by 1990, would utilities find it economic to replace conventional sources with their own solar arrays. This picture is now changing. Problems with finding sites for new conventional facilities, as well as climbing costs, are beginning to cause utilities to consider procuring photovoltaic units as an optional power source (Fig. 5–5). When a utility considers constructing a new generating plant of any sort, it is concerned with the cost of power from the new facility, not the cost from its combined existing generating capacity, most of which is bought and paid for.

The two most critical figures for a utility are the rate it is permitted to charge for the sale of power and the rate of return on investment it is permitted to earn.

Congress has already mandated, in the Public Utilities Regulatory Policy Act (PURPA) of 1978 (P.L. 95-617), that utilities' rate structures may not discriminate against small producers of electricity, nor may these rates force other customers, in effect, to subsidize the same small producer. If a utility decides to buy power from distributed photovoltaic sources, it must do so at a fair price—that is, it must pay the difference, in both its operating and its capital costs, that arises from having photovoltaic systems on the grid. High sellback rates, approaching 50 percent of the price of purchased electricity, will provide a strong boost for distributed photovoltaic systems, especially in the early 1980s. Sellback rates will be established by law or set by public utility commissions, in accordance with the PURPA mandate. The California State Energy Commission now requires utilities to buy back excess power from customers who have solar installations at 100 percent of the rate they are charging for power.

Thus far, no one knows how the utilities will respond to customer-owned photovoltaic systems. The utility industry is not monolithic, and differing situations could force differing responses. Some utilities have

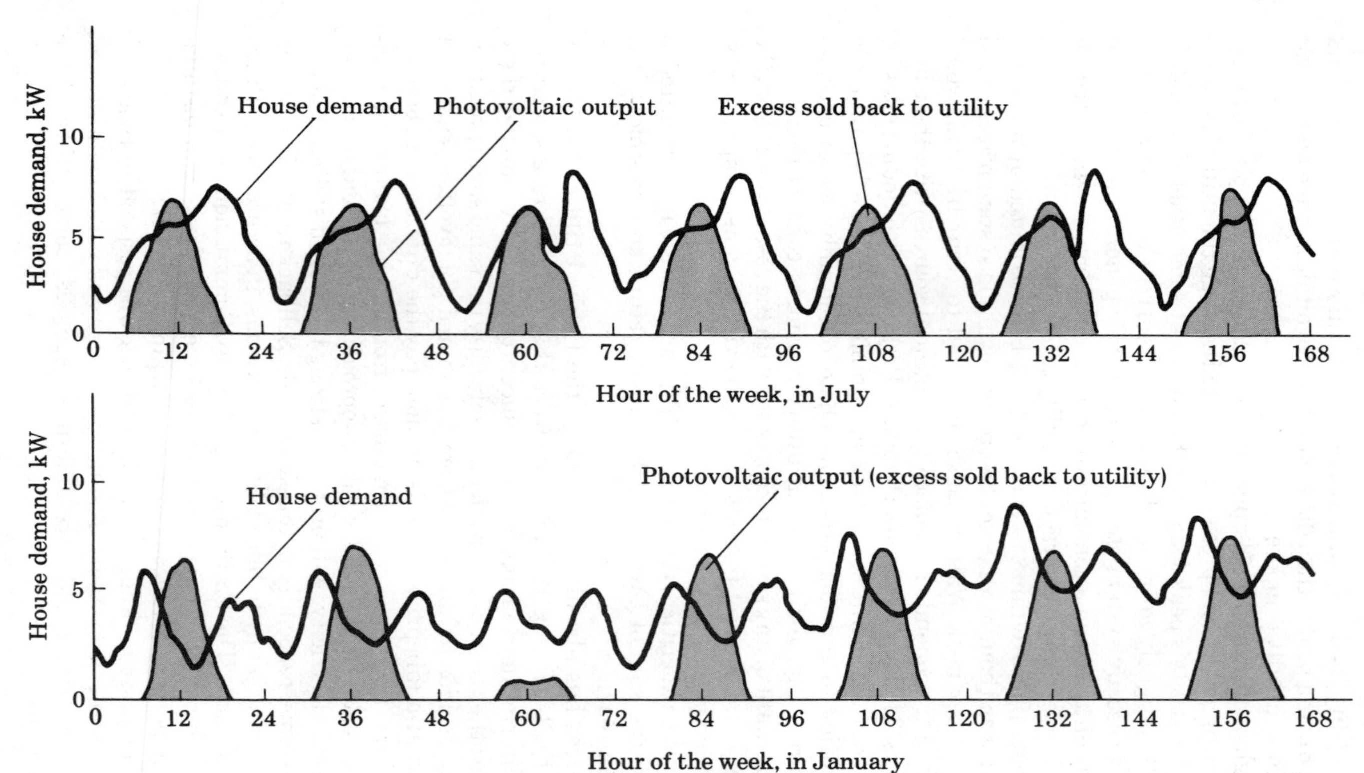

Fig. 5–7 Simulated residential electric power load and solar photovoltaics output (8 kWp system), Phoenix, Arizona. Peak midday power production from a solar photovoltaic unit does not usually coincide with household demand. A few hours storage capacity greatly increases the efficiency of the solar installation. Storage, though not essential, should be viewed as an integral part of the problem.

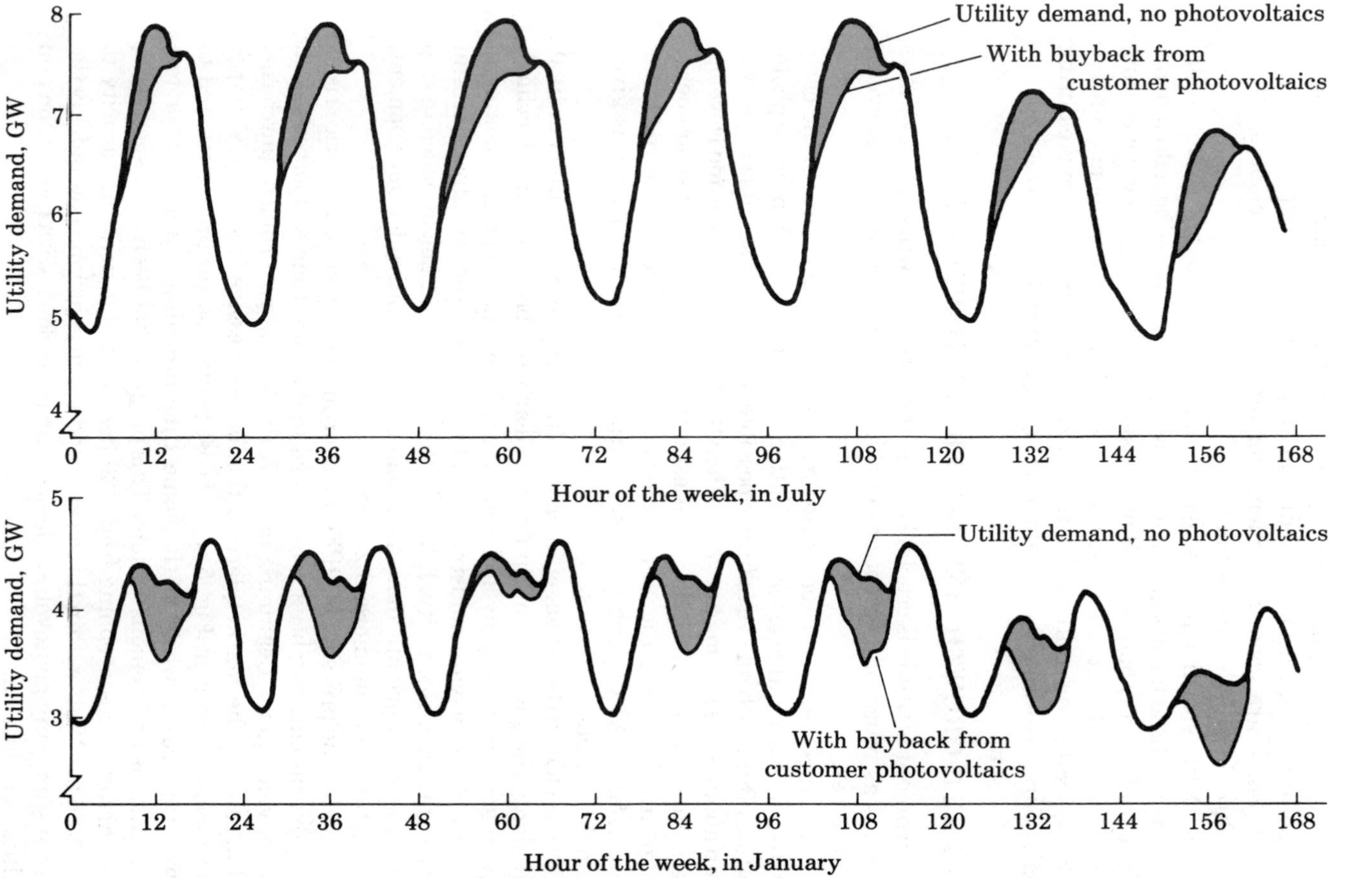

Fig. 5–8 Utility load profile with and without customer-owned photovoltaics for typical summer peaking utility, Phoenix, Arizona. From the utility's standpoint, buying power from customers who own photovoltaic systems can alleviate peak load requirements. A few hours of storage capacity would further ease peak loading.

trouble getting adequate fuel supplies; some do not. Some have trouble with pollution regulations or finding appropriate sites for expansion; some do not. Some are good investments for prospective stockholders; others have chronic trouble raising funds.

A utility's decision to resist or embrace distributed photovoltaics depends on these factors and more. Some utilities, foreseeing a drop in absolute profits, may oppose the competition offered by these new generating sources which do not fall under their direct physical and economic control. Others, seeing the contribution that distributed sources can make during peak hours, brownouts, and outages, may find much that is favorable.

MARKET GROWTH AND GOVERNMENT POLICY

Government attitude is crucial because the future growth of photovoltaics depends as much on perceptions of this technology as on market events themselves.

A glance back over recent history may be in order here. In the early 1970s, the National Science Foundation (NSF) began to support the photovoltaic effort with an eye to future possible terrestrial uses. Its budget was then about $1–2 million per year. When the Energy Research and Development Administration (ERDA) was established in 1974, most solar energy programs, including the NSF photovoltaic team, were transferred to the new agency. As the energy crisis sharpened, more attention was given each year to photovoltaics.

The DOE RD&D budget in photovoltaics rose to $157 million in 1980. All along, the government's intention clearly has been to help the technology to grow primarily by assisting research and development, looking to the day when private industry would find this an economically successful way to produce electricity. Likewise the government has intended to step out as soon as private industry assumes the initiative for development, production, and commercialization.

Under the pressure of OPEC oil shortages and consequent price rises, the government has been constrained to do more than support research and development. Beginning in 1977, the federal government purchased solar cells in the marketplace in the four successive "block buys" (large purchases), shown in Fig. 5–1, to boost private sector production and to lower unit costs, while at the same time providing equipment for test applications and demonstrations. This program has been quite successful, with array prices dropping from $30 per Watt of generating capacity to about $6 per Watt within three years and vividly showing how quickly the cost of supplying photovoltaic devices can come down with innovation and volume production.

Table 5–5 Estimated Total World Photovoltaic Market Related to Federal Government (DOE) Purchases (kW/yr)

	'74	'75	'76	'77	'78	'79	'80
Private Sales	100	200	404	660	1,000–1,200	1,600–2,000	3,000
DOE Purchases	0	0	46	90	200	500	1,000
Total	100	200	450	750	1,200–1,400	2,100–2,500	4,000
DOE % of Total	0	0	10	12	16	25	25

These federal purchases are summarized in Table 5–5, and it can be seen that they range from less than 5 percent to about 30 percent of the total market demand.

The federal government does have ample mechanisms to influence the course of a technology or an industry; these include direct grants, low-interest loans with or without government guarantees, direct procurements for government use, and various forms of subsidies to purchasers. In early 1980, photovoltaic systems were added to the classes of solar equipment for which a tax credit is allowed. The allowance was also raised to 40 percent of the first $10,000 for purchase or installation. Absolute amounts of further federal support for photovoltaics will depend on the perceived benefits to society. The mechanisms employed by the federal government will be based on many factors: the equities involved, ease of administration, possible time lags involved, and public perceptions of the acceptability and usefulness of such activities.

Congress has been a strong advocate of purchases under the Federal Photovoltaic Utilization Program (FPUP) which encouraged various agencies to buy and use photovoltaic systems. Uses have ranged from lighthouses to military housing to park lighting. In 1980 the largest single photovoltaic system in the world supported by federal assistance was a 100-kW system at Natural Bridges National Monument, Utah, which powers the park complex. The second largest system is a 60-kW system atop Mt. Laguna, California, which provides 10 percent of the daytime power used by an Air Force radar station there.

From a purely economic standpoint of initial cost, neither of these installations is economic; they nevertheless have been judged worthy from the standpoint of helping the photovoltaic industry to grow and providing it with operational experience. In 1980 the FPUP program, then three years old, was retired because the Department of Energy could use available funds more effectively and promptly for test applications under its direct control. Other federal agencies are encouraged to buy photovoltaic systems as they see fit.

Under FPUP, a total of 3,000 separate projects have been planned and funded. Collectively, they will generate 750 kilowatts at peak power. Most are small applications, typical projects involving water pumping, telecommunications, navigational aids, and power for residences for park rangers, the military, etc. By the end of 1980, 300 projects were completed. Another thousand will be in operation by the end of 1981. Locations extend to 45 states.

The record makes clear the value of government support of photovoltaics market development to the present time. It is arguable what form of support or inducement would be most effective and most equitable in the future: government purchases as in FPUP; subsidies to users, as in the tax credit approach; or subsidies to producers as through low-cost loans for equipment and fast tax write-offs, etc.[13] In any event, government support of R&D is expected to continue at significant levels.

In the place of the FPUP program, a form of open-ended government purchase of R&D has been initiated (*P*rocurement *R*esearch and *D*evelopment *A*nnouncement, PRDA). DOE has paid for competitive designs for a number of system evaluation and experiments, and then contracted further with the winners to construct the systems. Nine PRDA projects are now under construction.

DOE has itself recently issued a detailed analysis of how much federal money might be required to meet the Congressional goals spelled out in the Photovoltaic RD&D Act. These goals include doubling the nation's photovoltaic generating capacity every year for the years 1978–88, and providing a total installed photovoltaic capacity of 4 million kilowatts (4 gigawatts) by 1988. In addition, systems costs are projected to fall to $1 per Watt by 1988, with the nation's private sector purchasing 90 percent of output. The Act authorizes, subject to appropriations on an annual basis, $1.5 billion to accomplish these ends. DOE has examined a number of options, and concluded that within this $1.5 billion range, considerable progress could be made toward the goals of the Act. However, for assurance that the 1988 goals will be met, a budget of from $3 to $10 billion may be required. (By way of comparison, more than $100 billion has been spent to support development of commercial nuclear reactors.) Much of the additional support may come from private sector investments, and there are encouraging signs that this will happen. For this to occur, the government must be steadfast in its backing for the program.

Most of this money would go to building up production capacity by supporting early market demand. It is clear that the role of government in the initial phases of market expansion of photovoltaics is crucial.

---

[13] There are strong signals that the Reagan administration will sharply curtail market development activities by the government, putting available funds into R&D instead.

INTERNATIONAL MARKETS

Government policy clearly is also crucial in shaping the international market. The United States currently has a large number of multilateral and bilateral agreements concerning solar energy. Foreign sales are projected to account for half of intermediate-term (1982–1986) sales. DOE's International Photovoltaic Program Plan indicates that three major foreign intermediate-term markets are low-lift and medium-lift water pumping applications, and electric power systems for remote villages. The total size estimates for these markets range from 16 to 315 MWp/year. Foreign markets have a large potential for photovoltaic sales in the intermediate-term photovoltaic system price range of $3 to $10/Wp.[14] Another recent DOE report forecasting foreign market demands for photovoltaic systems arrived at similar conclusions indicating that foreign sales of communications and cathodic protection systems have the potential to grow more rapidly than U.S. sales in these markets during the early 1980s.[15] However, the United States' sales and foreign sales are projected to remain about equal until 1985, because of slower penetration in foreign markets. Many of these multilateral and bilateral agreements address research, development, and demonstrations of photovoltaic technology. The majority mentioning photovoltaics call for exchange of research and development information. For example, the U.S.-Israel agreement includes a research and development project on luminescent solar concentrators. The U.S.-U.S.S.R. agreement includes an exchange of technical information on gallium arsenide and multiple junction silicon cells. Such cooperative efforts may hasten maturity of photovoltaic technology.

Two important agreements include photovoltaic demonstrations. The largest project is the Saudi Solar Village Project to construct a 500-peak-kilowatt village power system to serve two villages near Riyadh, the capital of Saudi Arabia. The project will cost approximately $17 million; equipment is being procured from American firms.

In another agreement, a 20-peak-kilowatt experimental photovoltaic power system constructed beside a solar thermal experiment is part of a U.S.-Italy bilateral agreement. The U.S.-Italy agreement will also field a 5-kW photovoltaic village system modeled after the Schuchuli Indian Village in Arizona.

Until recently, it was thought that the developing countries would be the major market for photovoltaics in the early to mid-1980s. Electricity

---

[14] U.S. Department of Energy, *International Photovoltaic Program Plan: Report to the Congress in Accordance with the Solar Photovoltaic Energy Research, Development and Demonstration Act of 1978,* vol. 1 (Washington, D.C.: U.S. Department of Energy, 1979), p. 18.

[15] U.S. Department of Energy, *Export Potential for Photovoltaic Systems: Preliminary Report,* prepared by Battelle Pacific Northwest Laboratories, DOE/CS-0078 (Washington, D.C.: U.S. Department of Energy, 1979).

for water pumping is a prime need, and photovoltaics is well suited to this since most of the countries lie in the Sunbelt and have ample insolation. Much of the land is semi-arid, but usually there is ample water within perhaps 50 meters (150 feet) of the surface.

In addition, there are an estimated ten million villages in these countries, with populations of 100 to 1,000 persons each. An estimated two billion people—nearly half the world's population—live in these villages. They need a minimum amount of electricity for basic needs such as cooking, lighting, communications, crop grinding, and if possible for sewing and refrigeration.

Foreign sales are expected to grow steadily as the price of photovoltaics comes down. Even so, the United States' market is also expected to expand rapidly in the mid-1980s, and the domestic market may quickly outdistance the foreign market in the intermediate term, 1985–1990.

Principal purchasers in the foreign market are likely to be governments and international developers and the financing agencies that work with them.

HOW DO THINGS STAND NOW?

Despite foot-dragging by the Carter administration, the budget for photovoltaics was not reduced; it even kept slightly ahead of inflation. The net effect is that, during the seven years following the Arab oil embargo of 1973 and the Cherry Hill conference, a series of enlightened administrations and congresses has provided a level of support for photovoltaics that, while not extravagant, has been at least healthy, and the effects have been dramatic.

Clear evidence of this is that the private sector is now allocating large amounts of risk capital—$100 million in 1979–80—to photovoltaic development. New plants with two-shift capacity, totalling 10–25 megawatts per year, are being built. (Annual sales in 1980 were only about 4 megawatts—one-tenth of the production capacity now under construction.) The industry appears to be planning an expansive and expensive growth to the billion-dollar-per-year level by the latter 1980s.

Since photovoltaics first began to attract interest because of the nation's space program, costs per peak Watt have already been reduced by two orders of magnitude, a factor of one hundred, and are now within a factor of ten of being generally competitive with utility power sources in most parts of the United States.

The single most important influence on the infant photovoltaics industry between today and the "pot-of-gold" market when installed systems will be universally available for $2 per peak Watt or less, is sustained, steady growth. There is no technological magic involved; rather growth will come from public realization that, far from being magical or exotic, photovoltaics is a mature and practical technology as of today, which offers even more attractive benefits for tomorrow. Ultimately, public awareness of its benefits will provide the driving political and economic force needed to bring photovoltaics into widespread use.

CHAPTER 6
Societal Aspect

There is every sign that our society will increasingly choose electricity as the preferred form for delivering energy to the point of use. Electricity is clean. It does not have to be hauled about. It is extraordinarily versatile in use and can be converted easily into light, heat, or motion. It can be used with great precision—in a computer, a television set, or a surgical device. Moreover, the distribution system, representing an investment of $1 trillion, is already in place. In every sense, electrical energy is high quality energy.

In the United States, we now produce and use about 250 gigawatt-years of electricity each year. (A gigawatt-year is the amount provided by a 1,000 megawatt plant operating without interruption for one year. This would power a city of about one million persons.) Use of electricity is increasing at nearly 4 percent a year, and recent studies project that by year 2000, Americans could use from 400 to 800 gigawatt-years annually. A doubling of present use to 500 gigawatt years would appear to be a reasonable, if not conservative, estimate of yearly use by the end of the century.[1] Where the additional energy will come from is an open question, but it is entirely possible that photovoltaics could provide most of it.

It is important to recognize some singular aspects of massive photovol-

---

[1] Conservation efforts, including the gradual appearance of more efficient equipment, have lowered the growth rate somewhat. But most authorities still predict at least a doubling of electricity use in the U.S. by year 2000. Chauncey Starr, former president of the Electric Power Research Institute says that by 2000, demand will be 2 to 2½ times the 1980 level. He admits to a large shortfall in supply. "Choosing Our Energy Future," *EPRI Journal* 5 September 1980): 6–11.

See Joanne Omang, "Electricity Demand Keeps Climbing, But at Slower Rate," *Washington Post,* 24 August 1980, p. A–16; Edward Teller, *Energy from Heaven and Earth* (San Francisco: W.H. Freeman and Co., 1979), p. 292; *Statistical Abstract of the U.S.*, 1979, p. 607: Table No. 1025 shows usage increasing 5 percent per year or more in the period 1980–1990, for a gain of 65 percent for this decade alone. Also see "Electricity Growth: Part Trend, Part Cycle?" *EPRI Journal* 5 (September 1980): 18–22. The observed growth rate from 1972–78 is 3.62 percent per year, the long term rate (1966–78) is 4.28 percent per year.

taic deployment. For instance the cost of photovoltaics is almost entirely attributable to the creation of capital plant; it is a one-time front-loaded expense. There are no fuel costs, and maintenance costs are minimal. Most of the expense of the changeover to this new energy source would be borne by the private sector as a normal part of new construction costs, or in connection with the maintenance of existing structures. Such costs would in essence be attributable to higher quality of living. Consider: only a few decades ago, most new housing did not have central heating. Now central heating is taken for granted, and the same goes for electricity. Our economic system is able to provide these amenities and many, many more attributes of modern living. Some people earn their living providing them, and all of us enjoy the results. So it will be with photovoltaic solar energy.

The primary issues are whether technology can provide the capability at reasonable expense in terms of the human and material resources needed, and what the social and economic consequences of widespread use of photovoltaics may be.

MATERIAL AND RESOURCE CONSTRAINTS IN THE DEPLOYMENT OF PHOTOVOLTAICS

For any technology to expand rapidly into the marketplace, there must be an adequate supply of materials at costs reasonable to the purchaser, or the technology must be amenable to changes that bypass critical shortages.

In only a few instances are special materials for specific cell types in critical supply.[2] Polycrystalline thin film cells will require increased production of trichlorosilane. For amorphous silicon cells, indium is the critical material in present cell designs. Cadmium will be the pacing material for cadmium sulfide-copper sulfide cells, but proven world reserves of 700,000 metric tons indicate no shortage.

For gallium arsenide concentrator cells, availability of gallium, germanium, and arsenic may restrain large-scale production. But concentrators require small cell area, and such systems are useful only in regions of high direct insolation.

No material shortages are anticipated in any of the silicon designs, because there is enough lead time to provide needed production facilities.

---

[2] Raw and bulk materials requirements for large-scale use of photovoltaics have been extensively studied by Battelle Pacific Northwest Laboratories. See N.E. Carter, et al., *Some Potential Material Supply Constraints in the Deployment of Photovoltaic Solar Electric Systems*, PNL-2971 (Richland, Wash.: Battelle Pacific Northwest Laboratories [1978]), pp. 1–46; and R.L. Watts, et al., *The Evaluation of Critical Materials for Five Advanced Design Photovoltaic Cells with an Assessment of Indium and Gallium*, PNL-3319 (Richland, Wash.: Battelle Pacific Northwest Laboratories [1980] pp. 1–99.

Purified silicon will be required in greatly increased quantities. Several production processes are available, and two plants are under construction that will reduce the cost of solar-grade silicon from $65 per kilogram to $14 per kilogram.

The diversity of cell designs and concepts assures a wide choice of materials. Material availability and cost will be only two factors among many that determine which concepts find their way into the marketplace. The advent of thin film cells (less than 100 micrometers thick) and automated production processes will tend to make materials supply ever less critical. Further, no single cell design is likely to corner the market, and a mixture of cell types will further dilute any shortage.

Widespread use of photovoltaics will have its greatest impact on ordinary structural materials—iron, steel, aluminum, glass, and copper. Until advanced batteries come on the market, lead may be in increased demand. Indeed, of all the materials involved, it appears that glass will be the pacing item, since it will be the material most used as the weather surface for photovoltaic modules. A few years lead time for expanding production will prevent glass from becoming a bottleneck.

PHOTOVOLTAICS AND THE ENVIRONMENT

Photovoltaic cells are sealed units and no by-products are created in normal operation. However it is possible for a cell to fail in a way that produces a local hot spot. Under extraordinary conditions, short circuits could occur and cause sparks, with consequent risk of fire. Electrical protective devices are available and would be incorporated in systems to sense such abnormalities and shunt the current past the troubled part. In general, electrical safety can be assured through normal code practices regarding insulation, lightning protection, safety fusing and related matters. With reasonable attention to such aspects of system design and installation, photovoltaic systems should prove fully as acceptable from an operational standpoint as, for example, home television sets.

In the longer future, the partial shading of significant areas of land by massive photovoltaic arrays merits consideration. Shading can be greatly alleviated by the use of roof areas and south-facing walls for arrays. Increased cell efficiency will give most homeowners adequate array space on their dwellings. Moreover, land that is subject to partial shading can be used for many purposes. Horticulturists use this principle. In many areas of the world, additional shade would be a distinct benefit. The matter is not considered serious because the total land area needed in most countries is small compared to what is available.

From the standpoint of material hazards, there are three broad areas of concern: the release of toxic gases due to system malfunction; exposure of industry workers to toxic gases and dusts; and solid waste disposal, or release of gases, during photovoltaic material mining, manufacture, or disposal. All of the materials involved are well known to industry, and normal Occupation Safety and Health Administration (OSHA) practice should provide adequate safety. Most of the potential hazards can be mitigated with some research, normal OSHA practice, and genuine concern for potential problems.

Release of toxic gases during system operation and malfunction, for instance, could lead to the possibility of asphyxiation from the toxic products of material decomposition. Materials such as electrical insulation plastics have caused death in extensive fires involving electrical equipment other than photovoltaic systems. Some polymer materials used in concentrating optics (e.g., methyl methacrylate and vinyl chloride) release noxious fumes in a fire. Arsenic in gallium arsenide cells could vaporize in a fire and oxidize to a highly toxic arsenic trioxide. The decomposition and vaporization of cadmium sulfide at very high temperatures could lead to toxic fumes.

Exposure of workers in the photovoltaic industry to toxic gases and dusts will require monitoring and control as for any industry. The materials being considered for photovoltaic energy systems—silicon, gallium, arsenide, copper sulfide, copper oxide, tellurium compounds, etc.—are all already produced in major quantities by the $10 billion semiconductor industry which has a good record in meeting worker safety standards.

Approved health and safety codes from such industries as plastics, glass, and chemicals should be directly transferable to photovoltaic manufacturing operations. Photovoltaic cell manufacture presents no unique hazards, and should be much less hazardous than coal mining.

Other matters that will require attention include exposure to silicon dust in cell preparation, phosphine and boron trichloride in cell doping, and cadmium compounds during material production; inhalation of arsenic compounds during cell fabrication; high pressure release of arsenic trioxide in preparing gallium arsenide; and the handling of similar toxic agents. It will be important that toxicological research be expanded if new species and levels of toxic materials are generated. The industry, government agencies, and environmental researchers are beginning to perform these studies.

The disposal of mining and plant wastes derived from photovoltaic systems production will require normal monitoring and control as production is expanded. For example, gallium extraction yields mercury and large amounts of alumina sludge. The Environmental Protection Agency is now studying these wastes as they relate to containment at landfill sites.

Widespread application of any new energy technology can expose a large number of users and production workers to health hazards and create waste problems. However, photovoltaic systems are expected to enjoy extensive deployment without incurring exceptional environmental problems,[3] and the use of photovoltaics in lieu of expanded fossil and nuclear facilities will clearly be a net gain from the standpoint of protecting the environment. Dr. Louis G. Stang of the medical department at the Brookhaven National Laboratory, an authority on the environmental health aspects of photovoltaics, says: "The conclusion that can be drawn thus far is that photovoltaic technology is one of the cleanest of all energy options and certainly deserving of vigorous development."

EFFECT ON EMPLOYMENT

The employment potential of photovoltaics can be estimated rather simply. It takes 50 gigawatts (50 billion Watts) of photovoltaic generating capacity operating for one year to produce one quad of energy in a typical U.S. location of medium insolation. At a system cost of $2 per Watt, which will be reached in the latter 1980s, 50 gigawatts would have a price tag of $100 billion.

The labor cost of manufactured products like this is about one-fourth the sales price of the product, in this case $25 billion. Let's assume that the average wage for a worker in the industry is $20,000 per year. Then $25 billion divided by $20,000 per person-year gives 1,250,000 worker-years required to fabricate 50 gigawatts of photovoltaic systems. Installation and service could add perhaps 20 percent, or another 250,000, making a total of 1,500,000 worker years. This is the number of jobs required to produce the systems and install them and service them the first year. (The maintenance function would extend over the life of the installation, which could be 20 years.)

Widespread adoption of the various solar technologies would create an enormous number of jobs, according to the Council on Environmental Quality.[4] The following is a general indication of the distribution of skills

---

[3] U.S. Department of Energy (ERDA), *Solar Program Assessment: Environmental Factors—Photovoltaics*, ERDA 77-47/3 (Washington, D.C.: Government Printing Office, 1977); and *Environmental Development Plan for Photovoltaics*, DOE/EDP 031 (Springfield, Va.: National Technical Information Service, 1979).

[4] Council on Environmental Quality, *Solar Energy—Progress and Promise* (Washington, D.C.: Government Printing Office, 1978), pp. 2–3.

Type	*Percent*
Scientists, engineers, architects, and managers (about 2,000 professionals presently are working on photovoltaics)	7
Foremen, technicians, and semi-skilled production operators	27
Manufacture of piece parts	8
Sales and merchandising	5
Manufacture of power conditioning electrical products	8
Installers: electrical, roofing and plumbing trades people	18
Maintenance and service people	27
Total	100

likely to be needed:[5]

Before the end of the century, a labor force of several million people will be in new jobs, not jobs taken away from present energy industries. Coal miners and others working today will not lose their jobs to the photovoltaics industry. The *increase* in energy will be provided by people working in photovoltaic production facilities and on installations.

IMPACT ON MUNICIPALITIES

The adoption of photovoltaics will bring some special opportunities and also some new responsibilities for municipalities and other local jurisdictions. It will be most important to assure that installation, maintenance, and service organizations are qualified through training and are fair in their dealings with the public. As the technology comes into the marketplace, entrepreneurs will seek to establish themselves. Some will be better qualified than others, and local government and better business bureaus must prepare to protect the public.

Aesthetics may become a concern, and how local government will face this issue is an open question. It will be not so much a new issue as a new twist to an old one. Photovoltaic arrays will not be a problem on new construction because they will be roof or wall panels, or possibly imitation shingles, designed into the building. They will be removable so that they can be replaced in event of damage or when their conversion efficiency finally, after 20 years or more, declines to a level that replacement is economically advantageous.

Arrays added to existing structures may not be as visually harmonious, especially in the early years, but experience will teach architects and

[5] "Employment Impact of the Solar Transition," by Leonard S. Rodberg, Joint Economic Committee, U.S. Congress, 1979; Meg Schacter, "The Job Creation Potential of Solar Energy," Second Open Workshop on Solar Technology, Solar Energy Research Institute (Washington, D.C., October 1979).

installers how to protect basic building lines yet obtain adequate exposure to the sun. Arrays will soon appear on the market in the form of siding such as aluminum siding (and every bit as waterproof), or resembling shutters, and in all types of roofing patterns. This variety will help make photovoltaic arrays adaptable visually to almost any kind of structure.

It is no simple task to predict how homeowners will feel about owning photovoltaic systems. Arrays will have to be mounted in full view on their houses, and will need periodic hosing down and occasional repair. There will be associated control panels and perhaps a bank of storage batteries. Most of all, owners will have to pay for the system in advance, or through financing, trusting that the equipment will operate at least long enough to break even with the cost of power from a utility—a matter of several years. Every homeowner will not suddenly rush out to buy a photovoltaic set. Many will prefer to continue purchasing power from the utility.

Large arrays of solar cells may at first look out of place. One Washington official, recalling recent publicity accorded the proposed new radar-invisible bomber and the $20 billion which Congress voted for the synfuels effort, wryly remarked that the sure road to success for photovoltaics is to rename our modules syncells and make them invisible.

But soon photovoltaic arrays will become status symbols and people will not want them to be entirely inconspicuous. In the early days of television, the proud owner of a new TV set was happy to have an antenna anchored high on the chimney. Eventually, photovoltaics will assume the role of a necessity. Architects will be offered a new challenge in blending arrays into structures harmoniously.

In time, sun rights will become a concern, and local government will enact ordinances to protect the property owner from loss of direct sunshine on his or her array. The owner will be free to sell sun rights, that is, an easement. This aspect of law will gradually be thrashed out in the legislatures, municipalities, and in the courts.

Finally municipalities, especially larger urban areas, can look forward with optimism to reduced air pollution, especially when the electric auto appears on the scene.

PHOTOVOLTAICS AND NET ENERGY

One key measure of the effectiveness of a new energy option is how long must a renewable energy generator make electricity before the energy produced equals the energy required to manufacture the generator? This is called energy payback time. There is much confusion as to precisely how to do net energy analysis and, therefore, how to determine the payback times for various options.[6,7] Most previous analyses indicate that present photovoltaic technology has a payback time of four years. That is, it would take four years before a photovoltaic system can generate enough electricity to equal the energy needed to manufacture it.

However, new processes for photovoltaic production can lead to both lower costs and shorter energy payback times, typically less than one year. Some of the advanced films like amorphous silicon and cadmium telluride can have energy payback in a matter of weeks!

In order to determine the payback, the energy content of all raw materials must be added to the energy used in manufacturing, shipping, and installation. If payback times of less than one year can be obtained, then a rapid market penetration by a new option can occur without greatly affecting the total energy consumed. In a detailed analysis prepared by Solarex, the concept of the solar breeder was proposed. Simply stated, a solar electric generator would power a photovoltaic array production plant that has an annual output of arrays with a capacity that exceeds the energy needed to produce the solar generator and the raw material consumed. If this balance is obtained, the first year of operation will create photovoltaic generator products that can produce more energy than is required to manufacture them. This concept is now fully possible; such companies as Solarex are installing new processes that are both cost- and energy-effective. This approach could create the massive solar systems required without reliance on present energy sources.[8]

It should also be noted that once a photovoltaic generator with the ability to save one barrel of oil is installed, then that barrel can be saved again and again without any investment other than maintenance for the life of the installation, possibly 20 years or longer. That is, the energy capacity of photovoltaic systems is equivalent to having a proven reserve 20 times larger than the energy rating. If the nation had 10 quads of installed photovoltaics capacity, it would be like having 200 quads of proven energy reserves. In addition, it would be there at a guaranteed fixed price. This is the real value of a renewable energy option like photovoltaics with its quick (less than one year) energy payback time. In these days of rampant price increases for oil and other energy sources, it is comforting to have an installed photovoltaic energy system which will produce electricity at a fixed price for the life of the system.

---

[6] National Science Foundation—Stanford Workshop on Net Energy Analysis. Institute for Energy Studies, Stanford University, and TRW System Group. Palo Alto, Calif., August 1975.

[7] D.W. Fraley, C.L. McDonald, and N.E. Carter, "A Review of Issues and Applications of Net Energy Analysis." Pacific Northwest Laboratories, Richland, Wash., September 26, 1977.

[8] In computing energy payback time, allowance is usually made for the energy consumed in producing the raw materials used.

PHOTOVOLTAICS AND INFLATION

There can be no doubt that the repeated escalations in the price of foreign oil are the main cause of the current inflation. Former Treasury Secretary G. William Miller, also former chairman of the Federal Reserve, said that about one-third of our inflation is attributable directly to oil price escalation and the "indirect effects would be considerably more."[9] The perceived threat to oil supplies as well as to oil prices is edging up the prices demanded for all energy sources, regardless of kind or source. Because energy is a fundamental ingredient in everything our society makes or does, the shortage exacerbates inflation.

To combat inflation, we customarily raise interest rates and simultaneously reduce federal spending. These actions retard economic activity and cause unemployment to increase. Normally, the resulting shortage of money and cheaper labor cause prices to drop; deflation is achieved.

Eventually, the process must be reversed. To put people back to work requires the expenditure of energy. If energy is expensive, the cycle is quickly repeated. Worse, steadily rising energy costs may make it difficult to break the inflation process. Indeed, the economic phenomenon of the 1970s is that high inflation has continued even with high unemployment. For several years we have been locked in this economic corner, an enigma for classical economic theory.

Probably the best single step we could take is to find new sources of relatively cheap energy, especially sources that are not subject to the price manipulation of concentrated ownership. Such new energy sources should have two characteristics: highly diverse ownership, which means essentially free access to the supply; and abundant availability, which means the sources should be steadily renewed. Solar energy is obviously ideal in both regards. In fact, President Carter cited the development of solar energy as one means of combating inflation.[10]

Another critical link between inflation and the existing pattern of energy supply is the strong psychological element in the inflation phenomenon. Inflation feeds on the expectation of inflation. So long as we are tied to energy sources whose costs are spiralling, public awareness of this fact becomes a significant factor in making inflation worse. The price of oil is going up because the nations or the companies that own the supplies in the ground see these resources dwindling; they are determined to limit production and to ask ever more for what they release. The price of natural gas is similarly affected.

---

[9] ABC, "Issues and Answers," 29 June 1980. Interview of G. William Miller. Transcript from Tyler Business Services, Washington, D.C., 1980.
[10] President Jimmy Carter, State of the Union Address, January 23, 1979.

Because of the interchangeability of these energy materials, the price of coal will rise in response to the price hikes in other sources. In addition, the growing emphasis on environmental protection in both the mining and the burning of coal makes its use more costly. The price of nuclear power is similarly affected for precisely the same reasons.

Thus a major defect in all present major energy sources (hydroelectric power is a relatively small source) is that they directly contribute to the expectation of further inflation in prices. Because of the circumstances surrounding their production and use, oil, coal, natural gas, and nuclear energy may even be regarded as fundamental contributors to inflation. Because the supplies are obviously limited and being steadily depleted, they are inflationary energy sources. This perception, which has been so sharply reinforced in recent years, can only grow worse until we begin to transition to renewable sources.

Photovoltaics can be applied in ways that will relieve U.S. dependence on foreign oil. This should be the first objective. Certainly photovoltaics will not cause any reduction in the amount of coal being mined now, not within the lifetime of anyone now living. Photovoltaics *will* make it unnecessary to triple coal production, which some would now have us believe is the only practical alternative left.[11] Similarly, the advent of photovoltaics will not cause nuclear reactors now in operation to be abandoned. But photovoltaics can make it unnecessary to build more nuclear reactors, or to turn to the breeder reactor, which society is clearly loath to do. No matter how fast photovoltaics comes on, present energy supplies—other than foreign oil—and the jobs associated with them are not likely to be curtailed. It is the additional supplies of the future that now are at issue, and there photovoltaics will make a difference.

PHOTOVOLTAICS OVERSEAS

It would be derelict to forget the millions who live in poverty in other lands. It is to them that the oil price hikes have dealt the severest blow, as the Brandt Commission has pointed out.[12] In urban areas of developing countries, electricity costs in 1977 were $.50 per kilowatt hour. In rural areas power from diesel generator sets was $1 per kilowatt hour or more, and $12 per kilowatt hour from primary batteries. In the years since, matters have grown steadily worse. Photovoltaic technology can bring hope to many nations who can no longer afford to import oil for the 15 million diesel elec-

[11] Carrol L. Wilson, "Coal: Bridge to the Future," Report of the World Coal Study (Cambridge: Massachusetts Institute of Technology, 1980).

[12] The Brandt Commission was an independent international group headed by Willy Brandt, former chancellor of West Germany and holder of the Nobel Peace Prize.

tric power generators now in use.[13] The spillover effect from the application of the technology in developing countries could have beneficial long term political results potentially as important to the United States as solving the foreign oil problem today. Providing electrical power at the village level in the underdeveloped world can bring hope. It can also open the door to education, which is essential for beginning to deal with such seemingly intractable problems as excessive population and consumption growth rates.

But whether it should be or not, humanitarian concern is not the only motivation, perhaps not even the primary reason, for present interest in transferring this new technology into the developing world. From a pragmatic point of view, this developing technology needs to be applied in places where it now makes sense economically, in order to build up the production base and thereby bring prices down. If Village X is about to buy a diesel generator for the first time, or Village Y about to replace its worn out generating equipment, and a photovoltaic system is competitive, it is in both their interest and ours that they consider photovoltaics. Already, photovoltaic systems are directly competitive with diesel electric power in countries that must import oil, if the cost of oil over the life of the system is included! As the price of oil continues to climb and the price of photovoltaics continues to drop, the scales will steadily tip toward photovoltaics. By the mid-1980s, all diesel generator sets will be replaced with photovoltaics as they wear out (unless a country has its own indigenous oil supply). One of the world's largest manufacturers of diesel-electric generators has just decided to add a line of photovoltaic equipment to its offerings.[14]

A number of federal agencies are already involved in demonstrating photovoltaic systems in other nations and facilitating their widespread use. Activities have included both system experiments abroad and system field tests in the United States under conditions similar to those in developing nations. In addition, some resources already have been devoted to the development of balance-of-system components for use in international applications.

The Agency for International Development (AID) and DOE have been responsible for projects including the construction of either pumping applications or general village power photovoltaic systems. A village system for Tangaye in Upper Volta is working well; the grain grinding experiment has increased productivity so that the village is now a net seller of food. The use of photovoltaic electricity to pump fresh water, as well as for lighting, communication, and food preparation and refrigeration, can be a

---

[13] Private communication with Royal Dutch Shell.
[14] Private communication.

major factor in the lives of the several hundred million people who have no electricity.

A final word of caution is in order. With photovoltaic technology loosed on the world, the world will never be the same. History has shown that the ramifications of any new technology can be unexpected. But such speculation is hardly reason to hesitate. After all, life is a one-way street, and so is civilization. We have to move ahead one step at a time or we perish. By any measure, the present outlook for photovoltaics is all positive.

OTHER ISSUES

On energy matters, U.S. society is segmented in terms of ownership, uses, employment, living habits, even differences in climate from one region to another. These things pit different parts of society against each other in endless struggle. But ideologically there are two camps. Those persons and organizations that are users of energy are spearheaded by populist leaders who are committed to such social goals as a better physical environment and more equitable access to and control over energy sources and energy distribution enterprises. Confronting this group are the energy producing and supplying organizations, corporations, and their owners (and those who see matters as they do). The struggle between these two camps is nothing new. For years it has been punctuated by disputes and brawls until hardly an eyebrow is lifted over another contest about, say, public versus private ownership of an electrical utility, or by one more court test between "dams" versus "parks" groups.

Nevertheless, the energy crisis has intensified the usual interplay between the two encampments as never before. The march of events focuses unresolved issues in a way that produces severe strains on our whole socioeconomic system. For example, Amory Lovins, leading proponent of appropriate technology, says of the utilities that they are "very powerful and unenlightened. Even if under public ownership, they won't become benign. In Britain, they became less responsive . . . messing around with organizational form doesn't help."[15] He advocates energy technologies that will put control of energy supplies back in the hands of the people. This is an extreme view. Even in places such as Minnesota, where all electric utilities are publicly owned, the gap between the managers of central power systems at one end of the line and the users at the other sometimes seems unbridgeable. But the advent of photovoltaics offers hope that a new relationship will develop between utilities and their customers.

---

[15] "Hayden and Lovins Debate Institution Roles," *Solar Engineering*, October 1979, p. 24.

We are entering a truly fascinating era in the history of electric power generation and distribution. Since 1882 when Edison ran wires down the street from the world's first commercial central station generator in New York City, the trend has been all one way—toward bigger generators and longer lines. Photovoltaics offers the first real opportunity to reverse that trend.

The utiltities are now having serious problems keeping up with demand. The famous blackout of New York City on November 9, 1965 brought the matter to public attention in dramatic fashion. The energy crisis has added to utilities' woes. Not only must they meet the growing load, but do it in the face of fuel scarcities. Worst of all, spiralling fuel costs compel utilities to request repeated rate hikes, which further infuriates the public.

Michael Gent, executive director of the National Electric Reliability Council, points out that the utilities, which collectively burn 1.7 million barrels of oil a day, have been urged to scale this down to 1.5 million barrels per day by 1989.[16] Even this small cutback, he says, will take a prodigious effort. "The financial condition of the electric power industry is the worst it has ever been," says Gordon Carey, vice chairman of Commonwealth Edison of Chicago; this makes it virtually impossible to enlarge capacity or switch energy sources.[17] The utilities, it is clear, are caught in a blind alley.

Under these circumstances, photovoltaics could be a boon to the utilities as well as to the public. The utilities are offered an alternative power source that can be located anywhere in the system because it is modular in nature, has near-zero pollution, and is likely to be popular with the public. Homeowners can, if they wish, provide their own electricity with a photovoltaic system, and be freed from total dependence on the utility. Utilities are relieved from the steady pressure to provide new generating capacity and they can maintain earnings by buying power from homeowners during daylight hours at reduced rates and selling back at night at regular rates. Utilities can make money as distributors of power as well as by selling power they produce. From the public standpoint, the tremendous distribution network—the users paid for it—is an asset not to be scorned.

How the utilities and the public agencies that regulate them respond to photovoltaics will depend on their understanding of the technology and what it offers, and a willingness on all sides to try new approaches. The point is that the new photovoltaic technology will give us options that we have never had before—not since the dawn of commercial electric power. Lawmakers, regulators, utilities, and the consuming public will be chal-

---

[16] At a conference of energy specialists at Stanford University in June 1980; as reported by E. Marshall, "Planning for an Oil Cutoff," *Science* 209, 11 July 1980, p. 246.
[17] Marshall, "Planning for an Oil Cutoff," *Science.*

lenged to break out of traditional thinking and hidebound customs and seek new approaches that take advantage of the opportunities photovoltaics now offers.

An issue of social equity arises in the early stages of commercialization. Initially, the most likely residential purchaser is in a high income bracket. The 40 percent tax credit (to a maximum of $4,000) offers the wealthier members of society greater opportunity to procure photovoltaic systems. At first glance, this may appear grossly inequitable. But if someone must "break the ice," why not the more affluent? It is no different from any other new technology; the wealthy were the first to purchase home air conditioning, the private auto, electricity, and television. Someone must be first to buy photovoltaics so that the industry will grow and prices fall through mass production and technological advances. Whoever is the first to buy photovoltaics is saving oil or other fuels, reducing air pollution, and above all bringing photovoltaics one day closer for the rest of us.

In something so fundamental as energy, political authority can do little more than promote an attitude of reasonableness among the contending parties to assure that society continues to function smoothly while the people sort out their priorities in the marketplace, the political arena, and in their private lives. It is the people who finally must decide what energy future they want and why. If they are adequately informed, Americans will reach the right decisions and do so in a reasonably harmonious and equitable way.

THE TOTAL MARKET: HOW LARGE AND HOW SOON?

Will society adopt this new technology and apply it rapidly, or will photovoltaics be allowed to languish, coming slowly into use in some far-off future? What will be the size and timing of the total market? These questions follow a discussion of societal issues because too often the future of new ventures has been projected on the basis of purely economic factors with erroneous results. Nuclear power may be a preferred source of electricity on economic grounds, but the public has rejected it for other reasons. One wonders if coal may in time suffer a similar setback.

Concerns about market potential have spawned a series of studies on the future of photovoltaics. Unfortunately, each has narrowed on some special issue or aspect of the subject, or was motivated by some unique concern. The issue of total market size and timing has never been properly addressed. An extensive study by BDM[18] focussed on present uses

[18] BDM Corporation and Solarex Corporation, *Photovoltaic Power Systems: Market Identification and Analysis—Final Report,* vol. 1: *Executive Summary and Main Report,* prepared for U.S. Department of Energy, Office of Energy Technology, Division of Solar Energy, contract EG-77-C-01-2533 (McLean, Va.: BDM Corporation and Solarex Corporation).

Table 6–1 Estimates by the President's Council on Environmental Quality for Maximum Renewable Energy Sources Contribution to U.S. Energy Supply under Conditions of Accelerated Development (April 1978). Units represent quads of displaced fossil fuel per year.

Technology	*1977*	*2000**	*2020**
Heating and Cooling (Active and Passive)	Small	2–4	5–10
Thermal Electric	None	0–2	5–10
Intermediate Temperature Systems	None	2–5	5–15
Photovoltaics	Small	2–8	10–30
Biomass	1.3	3–5	5–10
Wind	Small	4–8	8–12
Hydropower	3	4–6	4–6
Ocean Thermal Energy Conversion	None	1–3	5–10

*The estimates in these columns are not strictly additive. The various solar electric technologies will be competing with one another, and their actual total contributions will be less than the sum of their individual contributions.

and near and intermediate term applications only, involving small units for use at isolated sites. It did not consider major applications to save energy. The SERI Venture Analysis[19] examined whether, and under what circumstances, a proposed $380 million, 8-year "market pull" initiative to promote production and reduce costs would be justified; a series of market scenarios assumed for purpose of analysis were of incidental import in themselves and admittedly involved great uncertainties. The content and objectives of earlier studies are summarized in the SERI effort.

In April 1978, the Council on Environmental Quality, a White House policy advisory group officially a part of the Executive Office of the President, produced its own assessment on solar energy.[20] The Coucil projected that photovoltaics could provide 2 to 8 quads per year by the year 2000, and as much as 10 to 30 quads per year by 2020, as indicated in Table 6–1. (One quad equates to 50 gigawatts of photovoltaic capacity operating for one year.) The high mark given photovoltaics reflected the confidence in its technical growth potential as well as in its basic simplicity and reliability.

A different kind of study was the Domestic Policy Review of Solar Energy conducted in 1978 at White House direction.[21] The DPR was a

[19] Dennis Costello, David Posner, Dennis Schiffel, James Doane, and Charles Bishop, *Photovoltaic Venture Analysis: Final Report,* 3 vols., SERI/TR-52-040 (Golden, Colo.: Solar Energy Research Institute, 1978).

[20] *Solar Energy—Progress and Promise,* Council on Environmental Quality (Washington, D.C.: Government Printing Office, 1978), p. 6.

[21] *Domestic Policy Review of Solar Energy.* Final Report: Research, Design and Development Panel, TID-28837, October 1978; and *Domestic Policy Review of Solar Energy,* TID-28834 (Springfield, Va.: National Technical Information Service, 1979).

thinly veiled effort by the Carter Administration to turn aside political pressure from a collection of solar enthusiasts. Various activist groups, which included serious students of the energy crisis as well as anti-nuclear, anti-big business, consumer protectionist, environmentalist, and similar elements, had found a common rallying point around the solar energy banner. Under the brilliant leadership of Denis Hayes, a highly publicized event called Sun Day was celebrated on May 3, 1978 to promote interest in solar energy. Speaking that day from the top of a mountain that is to become the home of the Solar Energy Research Institute on the western outskirts of Denver, President Carter announced that the DPR would be held under the aegis of the Secretary of Energy and 17 other department and agency heads. Whenever so many government units are deliberately included in a study, it is likely to be a stalling operation rather than an agent for change.

The DPR addressed all the renewable energy options, not photovoltaics alone. The outcome, announced more than a year later by the President at the unveiling of a solar installation on the roof of the White House,[22] called for 20 percent of the national energy consumption to come from solar (renewables) by year 2000 (19 quads of a projected total energy demand of 95 quads).[23] The portion foreseen to be provided by photovoltaics is widely quoted as 0.1 to 1.0 quad by 2000. But the lower limit was based on competing with oil at $25 per barrel. After the OPEC price rises announced at the Bali meeting in December 1980, oil was selling for $40 a barrel. Prices as high as $100 per barrel are now considered quite likely by 1985–90.

The upper projection of 1.0 quad would be realizable "within the framework of traditional Federal intervention . . . with a set of comprehensive and aggressive initiatives"; translated, this meant the present relatively modest level of support would continue.[24] The DPR projected 2.5 quads by 2000 as the "technical limit," but significant achievements in the subsequent two years indicate that this number may be appreciably exceeded.[25]

With an election in the offing in 1980, the Carter people could not afford to affront such a vocal political group as the Solar Lobby, which had become the spokesman for the solar interests. Neither did they want, in the

---

[22] President's Message on Solar Energy, The White House, Washington, D.C., June 20, 1979.
[23] See "Fact Sheet—The President's Message on Solar Energy," Office of the White House Press Secretary, June 20, 1979.
[24] *Domestic Policy Review of Solar Energy*, p. 47.
[25] The uncertainties in the projections were acknowledged: "It is not possible to forecast total energy demand or the future use of specific energy sources beyond the near-term with any certainty. A number of unpredictable factors such as the course of energy prices, the availability of competing fuels, future environmental standards, public attitudes, and the effects of government activity will affect the pattern of future energy use." *Domestic Policy Review of Solar Energy*, p. 47.

face of inflationary pressures, to be saddled with more expenditures for solar, which could do little to ease the energy crisis in the near term and nothing before the 1980 presidential election.

The appointment of Denis Hayes as Director of SERI, replacing the ailing scientist Paul Rappaport, was clearly a step in the right direction.[26] But 19 quads from renewables by 2000 sounded better than it turned out to be on careful reading. To begin with, it contained 4.2 quads already coming from hydro and biomass (2.4 and 1.8 quads, respectively in 1977). One additional quad would come from passive design of buildings. Of the increment, over half would be derived from hydro, biomass and in the industrial and agricultural areas. In the "maximum practical" case, less than 8 quads was assignable to active solar, including wind (1.7 quads) and ocean thermal (0.1 quad). Such a tepid projection, far from challenging the nation's technical, industrial and commercial communities, represented little more than a *laissez faire* approach to the future. Not surprisingly, the net result of the DPR was to provide the lowest common denominator of the assorted participants rather than a challenge to the nation. Amory Lovins called the DPR "a largely vacuous and uncreative exercise." Congressman Robert Drinan termed the Carter administration's position "Hypocrisy . . . characterized by glowing rhetoric on solar energy, but little meaningful action." The wide publicity given the DPR has left the public with the impression that photovoltaics has little to offer in this century.[27]

Still another study, completed in January 1979 under the auspices of the American Physical Society, concluded that the cost reductions required to make photovoltaics practical indeed are feasible. It examined only the central station, utility-operated application, however, and found that "It is unlikely that photovoltaics will contribute more than about 1% of U.S. electrical energy by the end of the century." While this would be one quad, central station applications are the farthest in the future. The APS study no doubt reinforced the impression left by the DPR that photovoltaics is at best a long term solar option.

A favorite technique in analytical studies of future market opportunities is to calculate the breakeven price for photovoltaics—the price of

---

[26] Hayes' appointment led Rappaport to complain bitterly that SERI had "been politicized." There is no question that SERI should be foremost a center of excellence, competent and objective in all the solar technologies and their applications, including the socio-economic aspects of their use. But by virtue of its position, and at a high level of rationalism, SERI has a solemn duty to advocate. Rappaport seemed unable to acknowledge such a role, even though SERI had come into being because of political action, including the enthusiastic support of scientists, himself one of the leaders, who were convinced that extensive solar applications were economically practical. Private discussion between Stirewalt and Rappaport at 14th IEEE Photovoltaic Specialists Conference, San Diego, Calif., January 7–10, 1980.

[27] For an excellent short review of recent solar policy, see Steven Ferrey, "Solar Eclipse: Our Bungled Energy Policy," *Saturday Review*, 3 March 1979, pp. 24–26. Ferrey served on the Domestic Policy Review.

equipment in $/Watt at which the cost of electricity from photovoltaics will be the same as the cost from the existing source with which it must compete. The date that this will occur is assumed to be the time that significant market growth will commence for that particular application. Such a clear point of equilibrium is at best hypothetical. In the real world, the break usually begins much sooner, when the price is still two or three times higher than required for direct competition. A complex set of factors motivates early buyers. Sophisticated analytical constructs often contain such inconspicuous yet crucial oversimplifications.

With the widespread commercialization of photovoltaics yet to come, it may be meaningful to examine present and near-term growth rates and ask what factors, characteristics and conditions are inherent in the technology that may deflect it from its present course—in effect, to explore the problem in the negative.

The goal of 4 peak gigawatts total cumulative production by 1988, set by the Solar Photovoltaic Energy RD&D Act of 1978, represents a conservative estimate based on a consensus of knowledgeable people in industry, government, and the research community. It assumes a continued doubling of production each year, which is an extension of the trend in recent years.

Such a growth rate, when plotted on semi-logarithmic paper, yields a straight line. This is depicted in Figure 6–1 in the portion of the curve extending to 1988. Such a growth pattern is typical of new industries or enterprises—even of living things, be it a plant or a child. But obviously exponential growth—in this case a doubling each year—cannot continue indefinitely. Sometime after 1988, limiting factors will begin to intervene and the curve will begin to tip over and level out. What are the factors that will limit growth?

Lessened demand for electricity, in contrast with other forms of energy, obviously is a possibility. However, both increasing population and the steady trend toward urbanization make electricity a preferred energy form, especially in densely populated areas. If adequate quantities can be provided, the use of electricity is expected at least to double by 2000. Whereas today we are forced to curb the use of electricity, photovoltaics will allow consumption to grow. The technology is not likely to be checked by reduced demand for electricity.

Some technologies are inherently growth-limited. But the photovoltaic phenomenon, because it occurs in an extraordinarily thin layer of material, and because a number of materials work quite well, is a fertile field for exploitation. The materials for the most part are plentiful and relatively inexpensive. While balance-of-system components are not subject to similar dramatic improvements, mass production and design simplifications will permit continued decline in system costs. The long-term growth potential is clearly not technology-limited.

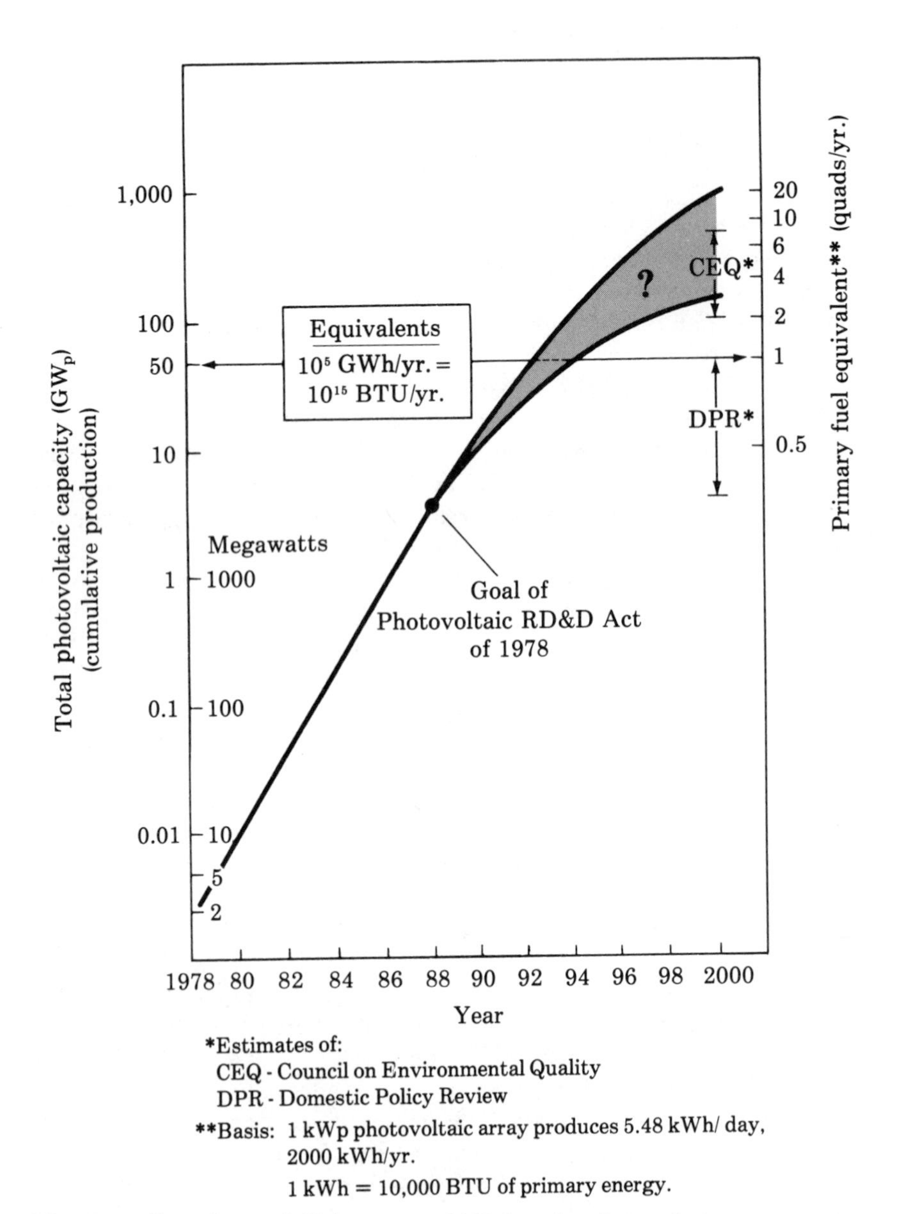

Fig. 6–1 Growth possibilities to year 2000 for solar photovoltaic energy systems. U.S. Production of photovoltaic equipment is expected to double each year through 1988, reaching a cumulative total of 4 GWp, the goal established in the Solar Photovoltaic Energy RD&D Act of 1978. Estimates of the Domestic Policy Review and the Council on Environmental Quality for cumulative production by the year 2000 are shown at upper right. No serious obstacles to growth are foreseen in the early 1990s, leading to the conclusion that cumulative production of 3 to 20 quads is possible by 2000.

Many technologies founder because very large units are required to make them economically attractive. Rapid scale-up in size is notorious for causing development delays, high failure rates and large cost overruns. In the solar field, the experimental Barstow "power tower" in California is a case in point; delays and cost overruns nearly killed it in 1980. But photovoltaic systems are about equally effective in any size. They are modular; additional units can be added like strings of lights on a Christmas tree. The risks inherent in extrapolating to very large units will not be an impediment in the application of the technology.

The necessary extensive manufacturing facilities *will* take time to construct and equip. While this will not be a bottleneck, the sheer numbers involved will serve to moderate growth rates in the 1990s.

Despite extensive automation, factories will require large numbers of skilled workers and trained supervisors. It will take time to develop this work force.

Retail sales, installation, and service organizations will have to be formed. Financing institutions will have to learn about photovoltaics. The gradual buildup of this supporting infrastructure will serve to temper the growth rate.

Institutional obstacles such as zoning regulations and construction codes that have not adapted to the technology will take time for adjustment.

In the long term, storage will become a significant factor in damping the exponential growth rate. A few hours of storage capacity for a residential system can be provided at reasonable cost today. But storage in the form of batteries or flywheels, while entirely feasible, requires components of significant size and weight. More efficient and cheaper batteries, flywheels, and other storage systems can be developed, but in general storage devices are not subject to miniaturization per unit of energy stored.

In many applications, such as water pumping and desalination, and in some industrial uses, no storage is needed. For others storage is a convenience or a way of making a photovoltaic system more efficient. In a few applications, storage is critical if an alternate source of electricity is not available. Storage will never be a bottleneck to photovoltaics, but in the mid- to latter 1990s, it will exert significant restraint on an otherwise explosive growth rate.

The ultimate limitation is market saturation. This will occur in stages as one market segment after another approaches full use of photovoltaics —say, 75 percent of total capacity. For example, new housing will be an explosive growth market in the 1984–90 period, but by the mid-nineties, practically all new housing will routinely be fitted with photovoltaics, and the *rate* of expansion into this field will level off. Similar effects will occur with commercial buildings, perhaps following housing by a few years.

But as one market segment is filled, another can open up. Photovoltaics may become the preferred energy source for processing synfuels from coal and oil shale. As things stand now, part of the coal must be burned to fuel the process. From one-half to one ton of coal must be consumed to convert another ton into other fuel forms. This inherent inefficiency is a major weakness of the synfuels effort. This process also intensifies the air pollution problem. Photovoltaics, probably as photovoltaic-thermal total energy systems, might render the synfuels venture considerably less unattractive.

The international market—the developing countries—now apparently the place where photovoltaics will first become economically competitive, may reappear as the large market of the future. It will be years before many peoples can afford even the low-cost systems that will eventually be available.

These limits to growth will begin to be felt in the 1990s, causing the growth curve to level out—it will continue upward but not as steeply. Cumulative production is expected to reach 150 to 1000 gigawatts (3 to 20 quads thermal equivalent) by 2000, as indicated in Figure 6–1. Equal installed operating capacity should lag this by not more than one year, as the bulk of production is near the end of the scale.

Dr. Harold Macomber, an international authority in the field, predicts that the industry globally will reach the $100 billion annual sales level before the year 2000. Cumulative production by that time, he thinks, will be 875 to 1000 gigawatts.[28] An expanded national commitment will probably be required to reach the upper level by that date. The actual level achieved will depend heavily upon public understanding, interest, and support.

The question of market size is important. Perceptions of market growth significantly influence achievement of widespread use. Perceptions of market potential influence the level of government support and they affect the commitment of funds by the private sector. Decisions hinge on estimates of future sales: the firms that make photovoltaic equipment, the distributors who sell it, and the banks that finance it will behave in accordance with their understanding of market growth rates and the absolute size of the market. Such expectations also have a vast unseen effect on public enthusiasm and support for the technology.

President Kennedy said that space is the new ocean and we would sail upon it. He set the goal of landing men on the moon and returning them

---

[28] Private communication with Dr. Harold Macomber.

safely to earth before the decade was out. the first moon landing was made eight years later. The challenge was backed by an unspoken yet profound expectation of fulfillment. Such prophecies tend to be self-fulfilling.

There are many aspects of society that photovoltaics will affect. This analysis has focused on the aspects that loom significant in the near term: materials supply, jobs, health effects, the role of utilities, and broader issues of social equity. No crippling flaw is seen in this technology. On the contrary, photovoltaics can satisfy many needs, such as transportation, that will be critical in the years ahead.

FURTHER READING

Brooks, Paul. *Speaking for Nature.* New York: Houghton Mifflin, 1980.

Carter, Luther J. "Environmentalists Seek New Strategies." *Science* 208, 2 May 1980, pp. 477–478.

Characterization and Assessment of Potential European and Japanese Competition in Photovoltaics. U.S. Department of Energy/SERI/Science Applications, Inc. SERI/TR-8251-1. Golden, Colo: Solar Energy Research Institute, 1979.

Courrier, Kathleen, ed. *Life After '80: Environmental Choices We Can Live With.* Andover, Mass.: Brick House, 1980.

D'Alessandro, Bill. "Photovoltaics: Milestones Passed, Miles Ahead." *Solar Age,* December 1979, pp. 12–14.

Environmental Development Plan for Photovoltaics. U.S. Department of Energy. DOE/EDP-0031. Springfield, Va.: National Technical Information Service, 1979.

Hayes, Denis. *Rays of Hope.* New York: W.W. Norton, 1977.

Ihara, Randal H., and Reid, Herbert G. "Technology, Democracy, and Solar Energy: The Moral Ambiguity of Solar Energy." *Proceedings, Helios: From Myth to Solar Energy,* 16–18 March 1978. Compiled by M.E. Grenander. Albany, N.Y.: State University of New York, Institute for Humanistic Studies, 1978.

International Photovoltaic Program Plan. 2 vols. Washington, D.C.: U.S. Department of Energy, 1979. Vol. I: *Report to the Congress in Accordance with the Solar Photovoltaic Energy Research, Development, and Demonstration Act of 1978.* Vol. II: *Appendix.*

[Jackson, Bruce]. "Utilities: A Growing Solar Program." *EPRI Journal* 4 (December 1979): 29–30.

Kidder, Tracy. "The Future of the Photovoltaic Cell." *Atlantic,* May 1980, pp. 68–76.

Lönnroth, M.; Johannson, T.B.; and Steen, P. "Sweden Beyond Oil: Nuclear Commitments and Solar Options." *Science* 208, 9 May 1980, pp. 557–563.

Marvin, H.H. "Solar Energy in Our Future." *Proceedings, Helios: From Myth to Solar Energy,* 16–18 March 1978. Vol 2. Compiled by M.E. Grenander. Albany, N.Y.: State University of New York, Institute for Humanistic Studies, 1978, pp. 304–314.

Morgan, Robert P. "UNIDO and Appropriate Industrial Technology." editorial. *Science* 203, 2 March 1979, p. 835.

"Net Energy in Photovoltaic Array Manufacture." Solarex Corp., Rockville, Md., 1978.

Partrain, L.D. "Six Kilowatt Residential Photovoltaic Power Systems Study: Design, Performance, Economics, Market Potential." Lawrence Livermore National Laboratory, Livermore, Calif., 1980.

Public Opinion on Environmental Issues. Council on Environmental Quality, *et al.* Washington, D.C.: Government Printing Office, 1980.

Reducing U.S. Oil Vulnerability—Energy Policy for the 1980's. U.S. Department of Energy. DOE/PE-0021: Washington, D.C.: Government Printing Office, 1980.

Solar Photovoltaic Energy Conversion. New York: American Physical Society, 1979.

Weiss, Charles, Jr. "Mobilizing Technology for Developing Countries." *Science* 203, 16 March 1979, pp. 1083–1089.

CHAPTER 7
A Possibility

The potential of photovoltaics is clear. The issue is not whether we as a nation will be able to obtain electricity directly from sunlight at prices that will be acceptable for everyday use. With continued modest support by the government, the breakeven point at which photovoltaics becomes economic for all uses will, as we have suggested, probably commence by the mid-1980s. The incursion of photovoltaics into the energy market will be very gradual at first and then accelerate; by 1990 a fundamental restructuring of the market will be underway.

The serious questions to be faced now are whether to make this happen sooner, and how to focus this new technology to alleviate present and impending national energy problems.

As we turn away from an era of energy gluttony, more efficient use of energy offers great benefits. Efficient use of energy involves changing both the equipment that uses energy—buildings, their insulation, and their heating and cooling plants; the machinery of transportation; and industrial processes and equipment—and the habits and mores of the society that uses energy. Both aspects deserve careful attention.[1] While we are still at our most wasteful, significant progress can be made rather easily, but in the long term, every new achievement will become more laborious and more expensive. The unit cost of conservation in terms of physical plant and social amenities (quality of life) can only go up. As one realist put it, "Conservation may be absolutely necessary as a tactic, but it is potentially disastrous as a strategy."[2] Conservation makes good sense especially while we are still at our most careless and wasteful, but it must

---

[1] Luther J. Carter, "Academy Energy Report Stresses Conservation," *Science* 207, 25 January 1980, p. 385; Eric Hirst and Bruce Hannon, "Effects of Energy Conservation in Residential and Commercial Buildings," *Science* 205, 17 August 1979, p. 656.

[2] Conference of energy specialists at Stanford University, June 1980, as reported in *Science* 209, 11 July 1980, p. 246.

be supported by vigorous activity to assure that adequate energy is available from acceptable sources at reasonable prices.

A new administration came to Washington in 1981 dedicated to the production of more energy. "The government has acted on the principle that the way to deal with energy is to do away with it," wrote President Reagan's energy policy task force. "Instead of unleashing the resources of a wealthy nation, we have, in the name of saving energy for some unspecified future time, tucked energy away like a rare bottle of wine . . . We reject the notion that the energy dilemma can be solved only by halting the use of energy . . . Energy resources are valuable only if they are produced and consumed."[3] It is clear that the emphasis is now on production rather than conservation.

The potential contribution of other energy technologies to the national energy supply affects the speed with which photovoltaics is commercialized. Research and exploratory development on each energy technology should continue at a level reasonably commensurate with its promise. Nuclear fusion work should continue until it is clear that further effort would be futile, or until the end of the tunnel is in sight. Let us drill enough deep wells to remove all doubt about the potential of geo-pressurized gas. Let us bulldoze over a few more counties here and there until we have revealed fully the implications of massive strip mining and synfuel operations.

The appropriate test is whether technologies such as these make sense for the long run, and if not, whether they will divert us from other technologies that are better matched to foreseeable future conditions.

NATURAL GAS

The U.S. supply of natural gas is limited and has been greatly depleted over the decades. One hope has been that present sources would be greatly augmented by geo-pressurized gas—large amounts of methane dissolved in deep underground waters primarily along the Gulf Coast. Recent evidence, however, has raised serious question about the amount that may be recoverable.[4] The latest drilling indicates that the methane is too dilute, and contained in too many small reservoirs to obtain it in quantity, even if the price of natural gas should double. Informed opinion holds that the dis-

---

[3] Report of the Energy Policy Task Force, Houston, November 5, 1980, Michael T. Halbouty, Chairman. See "Reagan Advisers Urge Phase-Out of Oil Profit Tax," *Washington Post,* 13 December 1980, p. A-1.

[4] Richard A. Kerr, "Geopressured Energy Fighting Uphill Battle," *Science* 207, 28 March 1980, p. 1455: "It now appears that even modest hopes for geothermal energy will be difficult to fulfill."

covery of new gas will barely maintain production at present levels and in any event, it is not an easy substitute for liquid fuels required in transportation. Synthetic natural gas similarly appears unpromising, at least for another decade.

COAL

In discussions of the nation's energy future, attention returns repeatedly to the major fossil energy resource, coal, which is in ample supply in this country.

The problems of mining and burning ever larger amounts of coal are becoming increasingly apparent.[5] In recent years there have been about 150 fatalities annually in the mines, and long term health hazards to workers and to the public at large are just now becoming fully evident.

Moreover, most of the coal must be hauled overland closer to the centers of population, where it is burned to create electricity or to provide direct heat. Transportation is becoming an increasing nuisance, especially for small towns in the Midwest through which dozens of trains pass daily.

If used for heat, coal must be burned at the point of use, increasing pollutants in populated areas. When coal is burned to make electricity, two-thirds of the heat energy is lost—which is why some have decried the steady trend toward expanded use of electricity.[6]

One major study of coal suggests that global usage may have to triple to meet world energy needs to the year 2000.[7] The Council on Environmental Quality says high coal usage could mean that 2,000 square miles of land would be stripmined in the rest of this century and that another 5,000 square miles, an area the size of Connecticut, would suffer subsidence. When the earth sinks as abandoned underground mines collapse, anything on the surface can be destroyed.[8] It is probably possible to manage, albeit with considerable discomfort, the mining and hauling of the vast quantities involved, but the core issue is what to do about the products of combustion, which weigh more than the coal. For every ton of

---

[5] S.C. Morris, P.D. Moskowitz, W.A. Sevian, S. Silverstein, and L.D. Hamilton, "Coal Conversion Technologies: Some Health and Environmental Effects," *Science* 206, 9 November 1979, p. 654.

[6] Amory B. Lovins, "Energy Strategy: The Road Not Taken?" *Foreign Affairs,* October 1976, pp. 65–96. Lovins rests his case on the high cost of new coal and nuclear central station plants compared to the cost of energy produced and consumed locally.

[7] Carroll L. Wilson, "Coal: Bridge to the Future," Report of the World Coal Study (Cambridge: Massachusetts Institute of Technology, 1980).

[8] Council on Environmental Quality, *The Good News About Energy* (Washington, D.C.: Government Printing Office, 1979), pp. 23–25.

coal burned, nearly three tons of combustion products result (mainly carbon dioxide).[9]

Of most immediate concern are the oxides of sulfur and nitrogen and particulate matter.[10] The larger particulates are relatively easy to remove at the stack, but the microscopic particles are disseminated for hundreds of miles, with long term implications for human health. Worse, sulfur and nitrogen oxides combine with water vapor in the air to form some of the strongest acids known. In recent years, higher stacks have been built to reduce local pollution, but this has only transferred the problem elsewhere. Increased coal burning in the Midwest is now producing acid rain in the Northeast.[11] Since the mid-1970s, fish have simply disappeared from thousands of lakes in New England and Canada. Indeed, acid rain has now become a matter of very serious contention in our relations with Canada, and the problem grows steadily worse.[12] Automobile finish is now being damaged by acid rain in some areas of the Northeast, causing car manufacturers to provide owners with special instructions for protecting their cars. Acid rain from burning coal now threatens Northern Europe as well. It is possible to remove much of the acidic components of stack gases at considerable expense, but this raises the cost of electricity and is therefore economically and politically unattractive.

The major long term issue that results from the burning of any fossil fuel—the build up of carbon dioxide (CO_2) in the atmosphere—is unresolved. CO_2 is highly volatile and there is no way it can be removed from the atmosphere on the massive scale involved. CO_2 in the atmosphere absorbs sunlight in several parts of the sun's spectrum. The earth must get rid of all the heat it receives from the sun, or it would get hotter and hotter. The earth reradiates much of the energy into space at the longer wavelengths. CO_2 lets in most of the sun's energy, which arrives in shorter wavelengths, but blocks some of the heat that the earth must get rid of. (This phenomenon is commonly known as the greenhouse effect.)

It has been estimated that a doubling of the CO_2 content of the

---

[9] U.S. Department of Energy, *Technology Characteristics—Environmental Information Handbook.* DOE/EV-0072 (Springfield, Va.: National Technical Information Service 1980), p. A-3. Appendix A is an excellent summary of the environmental factors involved in operating a coal-fired power plant.

[10] William S. Cleveland and T.E. Graedel, "Photochemical Air Pollution in the Northeast United States," *Science* 204, 22 June 1979, p. 1273.

[11] [Jenny Hopkinson], "Tracking the Clues to Acid Rain," *EPRI Journal* 4 (September 1979): 20.

[12] Bill Richards, "U.S. Conversion to Coal Would Add to 'Acid Rain' in Canada," *Washington Post,* 2 February 1980; Joanne Omang, "Acid Rain: Push Toward Coal Makes Global Pollution Worse," *Washington Post,* 12 December 1979; "Rain Falls Everywhere," *Washington Post,* 7 December 1980, editorial.

atmosphere will lead to an average global temperature increase of 4.5°F.[13] Since the beginning of the industrial revolution, the CO_2 content of the atmosphere is estimated to have risen 13 percent. At present rates of fossil fuel consumption, it will increase another 12 percent in the next 20 years alone. The effects of a change of even one or two degrees on the weather could be severe. For example, it is not unimaginable that the polar ice caps could begin to melt, causing oceanic flooding worldwide. The warming would be greater at the poles than in the tropics. In the last two decades, the CO_2 content of the air at the South Pole has increased 5 percent.[14]

Another complication is that the oceans absorb carbon dioxide from the air, and the effect that significant changes in the CO_2 content of the oceans would have on marine ecosystems is unknown. Plant life does use CO_2, but vast increases in forests would be necessary to counter significantly the buildup in the atmosphere. At the moment, economic expansion is causing the world's forests to be gobbled up at an alarming rate. There are still large areas of uncertainty about the "carbon budget."[15] Philip Abelson, editor of *Science,* writes about the CO_2 issue:

> Humanity is in the process of conducting a great global experiment. If unpleasant side effects are encountered, they cannot be quickly reversed.[16]

SYNFUELS

For a few months in 1979, the magic word that signaled an end to the fuel shortage was "synfuels." Synthetic fuels, or synfuels, is a generic term that includes both gas and liquid fuels made from coal as well as oil from oil shale. We would strip mine coal that lies near the surface in several western states, crunch it, and pressure cook it in closed retorts to produce an oily liquid akin to petroleum. This could be refined into gasoline and other liquid fuels. (In one fantastic scheme, the burning and distillation would be done *in situ* underground).

Because CO_2 is released both in the production process and later when the fuel is burned, the total CO_2 produced, and of course the energy released as heat, is at least 1.4 times greater, per unit of useful energy, than in the direct use of coal. The only valid reason for tolerating this

---

[13] Philip H. Abelson, "Energy and Climate," *Science* 197, 2 September 1977, editorial.
[14] Luther J. Carter, "A Warning on Synfuels, CO_2, and the Weather," *Science* 205, 27 July 1979, p. 376; Nicholas Wade, *Science* 206, 23 November 1979, p. 912.
[15] W.S. Broecker, T. Takashi, H.J. Simpson, and T.H. Peng, "Fate of Fossil Fuel Carbon Dioxide and the Global Carbon Budget," *Science* 206, 26 October 1979, p. 409
[16] Philip Abelson, "Energy and Climate," *Science* 197, 2 September 1977, editorial.

waste is the need for liquid fuels for transportation. However, because of the high costs and other problems, such as the use of water in states where water is scarce, the synfuels program is not expected to yield significant results before 1990.[17]

NUCLEAR FISSION

For the last two decades, nuclear power has repeatedly been urged as a satisfactory alternative to fossil fuels for generating electricity. Beginning in the 1960s, the utilities began seriously to turn toward nuclear. By 1980 there were some 90 reactors in operation with 100 more under construction or planned. But nuclear technology has turned out to possess an Achilles heel few had anticipated, a shortcoming probably more sociological than physical, but none the less real. People in substantial numbers have simply rebelled against nuclear reactors largely because they fear the effects of potentially uncontrolled radiation. The repeated demonstrations at Seabrook and the accident at Three Mile Island (TMI) are landmarks, symbolic of the issues involved.[18]

TMI did more than produce a psychological shockwave. Pennsylvania authorities have refused to allow Metropolitan Edison to charge its customers for the costly accident through the mechanism of the "fuel adjustment clause," which heretofore permitted utilities to bill their customers for substitute power when a nuclear facility is down, while continuing to recover from them the cost of the new plant. There is a real possibility that investment money for new nuclear plants may simply dry up.[19]

There are more problems in prospect. At present, nuclear plants store their spent fuel rods on the premises in tanks of water. Before the mid-1980s, the older plants may have to discontinue operation unless either more storage is provided or chemical reprocessing plants are set up to dissolve the rods in acid, separate and concentrate the fission products, and consign them to some "permanent" burial ground. Because the products of nuclear fission will be radioactive for several hundred thousand years,

---

[17] Eliot Marshall, "Synfuels Program Born in Confusion," *Science* 205, 28 September 1979, p. 1356; Ann Hughey, "Synfuels Versus the Political Realities," *Forbes*, 20 August 1979, p. 41; Kathleen K. Wiegner, "The Water Crisis: It's Almost Here," *Forbes*, 20 August 1979, p. 56; [Nadine Lihach], "Water Water Everywhere But . . . " *EPRI Journal* 4 (October 1979): 7.

[18] Richard L. Meehan, "Nuclear Safety: Is Scientific Literacy the Answer?" *Science* 204, 11 May 1979, p. 571; Eliot Marshall, "Assessing the Damage at TMI," *Science* 204, 11 May 1979, p. 594; Amory B. Lovins, "Invited Testimony for Hearings on the Costs of Nuclear Power Before the Environment, Energy, and Natural Resources Subcommittee of the Committee on Government Operations, U.S. House of Representatives," 21 September 1977 (photocopied).

[19] Richard Morgan, "Nuclear Power's Costs," *New York Times*, 5 December 1979.

scientists debate whether safe disposal is indeed possible. Meanwhile, the stockpile of nuclear debris continues to accumulate.

The nuclear breeder reactor, which would produce as much new fissionable material as it consumes, thus providing an open-ended energy source, continues in limbo for two reasons. In the hands of many nations, breeders, which produce plutonium, would be likely to contribute to the proliferation of nuclear weapons. In addition, plutonium itself is extraordinarily dangerous from a health standpoint. For these reasons, in addition to fear of nuclear reactors in general, civilized society is most reluctant to commit itself to long term dependence on plutonium as its crucial energy lifeblood.[20]

NUCLEAR FUSION

There is always the possibility of a breakthrough in nuclear fusion, the "clean" process in which heavy hydrogen (deuterium) is caused to react at extremely high temperature and pressure, forming helium and releasing energy in the process. Despite prodigious research, success in practical terms appears at least several decades away. There would still be problems with radioactive materials, as the equipment and containment housing become extremely radioactive. Moreover, fusion would entail a further commitment to the central station approach to electricity generation and distribution. The consensus is that nuclear fusion on a practical scale is unlikely until well into the next century. The Harvard study concluded that "In the United States there is simply no reasonable possibility for 'massive contributions' from nuclear power for at least the rest of the twentieth century."[21]

OIL

As everyone knows full well, the crux of the energy problem is imported oil. The United States' economy is being bled white to pay for the petroleum bought from abroad to the tune of some $60 billion in 1979. Even with continued curtailment in use, the price rose to $78.6 billion in 1980 for about 6.6 million gallons a day, and is expected to be $100 billion for

[20] "Political Problem," *Scientific American,* May 1980, pp. 78–82.
[21] Robert Staubaugh and Daniel Yergin, eds., *Energy Future: The Report of the Energy Project at the Harvard Business School* (New York: Random House, 1979), p. 135.

1981.[22] The effect on our economy could not be more pronounced if we manufactured and gave away the equivalent worth of automobiles, houses, computers, groceries and other goods every year. The annual trauma over a few billion dollars in foreign aid is a mockery by comparison. Each successive hike in oil prices is quickly translated into a steep increase in all other prices, and it is next to impossible to wring the resulting inflation out later.[23]

Abroad, our heavy dependence on foreign oil is a source of friction with our close allies, many of whom are more dependent on foreign oil imports than we are. Rising oil prices keep the developing countries in turmoil since they are being priced out of the market for the oil they need for bare subsistence. Human concerns aside, we obtain raw materials from many of these countries and they are a market for the products of our agriculture and technology.

There is no longer doubt that the end of the era of oil plenty is in sight. Indeed, the first oil crisis of 1973 was precipitated when it suddenly became clear to the major oil exporting nations that this was the case. The sudden jump in U.S. oil imports in 1973, as indicated in Table 7–1, caused the oil exporters to panic. Yet, despite the oil embargo of 1973, the repeated tightening of the noose by OPEC, and the steady escalation of prices, the U.S. managed to quintuple imports from 1970 to 1977.

Indeed, the oil embargo of 1973 was not so much politically inspired as the result of shock at the jump in American oil imports that year, coupled with the prospect of further rapid increases, as indicated in the second column of Table 7–1. Not until 1978 did we begin to turn the tide on imports. Meanwhile, total oil usage continues to rise. If the downward trend begun in 1979 is to be other than temporary, ever more stringent measures will be required. To be sure, rising prices will cause people to use less gasoline, and more efficient cars will ease matters for the moment. But at best these are stopgap measures against a trend that is long term. It is even being suggested that the private automobile may be headed toward a major decline.[24] And the staggering cost of imported oil is not the only issue:

> The important role that control of the oil fields played in the Shah's downfall meant that the oil fields in other countries would become, even more so than

---

[22] Hobart Rowen, *Washington Post,* 23 December 1978, citing estimates by U.S. Treasury officials.

[23] G. William Miller, former treasury secretary and former chairman of the Federal Reserve Board, says that about one-third of our inflation is attributable directly to oil price escalation and the "indirect effects would be considerably more."

[24] Lester R. Brown, *Running on Empty: The Future of the Automobile in an Oil Short World* (New York: W.W. Norton, 1979).

Table 7–1 U.S. Crude Petroleum Usage (million barrels per year)*

Year	*Imports*	*Domestic Production*	*Total*	*Imports as Percent of Total*
1970	483	3517	4000	12
1971	613	3454	4067	15
1972	811	3455	4266	19
1973	1184	3361	4545	26
1974	1269	3203	4472	28
1975	1498	3057	4555	33
1976	1935	2976	4911	39
1977	2414	3009	5382	45
1978	2275	3178	5483	42

* *U.S. Statistical Abstracts,* 1979, p. 759, Table No. 1329.

> before, a prime target of political dissidents. A revival of Islamic fundamentalism, co-existing with radicalism, will ensure a rich supply of political dissidents in Iraq, the Gulf sheikdoms, even Saudi Arabia, the mother lode of oil on which the Western world depends for political moderation and increased supplies.[25]

Our nation is hostage to the OPEC world, much of which is politically unstable.

It is difficult to overstate the degree to which this renders us vulnerable. The Congressional Budget Office estimates that a conceivable supply cutoff might well reduce oil imports by 3½ million barrels a day and last for a full year. This would cause a $270 billion loss to the general economy, an increase of 2 percent in unemployment, and a 20 percent jump in inflation.[26]

Continued dependence on foreign oil works directly against all our efforts to build a more secure and prosperous world order. For this reason, every energy proposal should be tested against one overriding criterion: how much and how quickly will it reduce foreign oil imports?

In summary, then, oil and natural gas simply cannot be discovered fast enough to expand the day-to-day supply. In real terms, world production of both may start to decline within a few years. Artificial substitutes in meaningful quantity are not in the offing. Use of nuclear energy is unlikely to increase greatly, certainly not in time to ease current problems. Both the near term and the long range consequences of greatly increased coal production and combustion appear unacceptable. Through

[25] Stobaugh and Yergin, *Energy Future,* pp. 30–31.
[26] "Oil Embargo Might Cost $270 Billion, CBO Reports," *Science* 208, 6 June 1980, p. 1123.

stringent conservation, the fuels we do use can be stretched farther, but the cost and difficulty of conservation increase with each new round.

TIME FOR A CHANGE

This is a sobering list of shortcomings, external costs and Faustian bargains. In the face of these realities, it is only prudent to ask whether a major commitment to solar photovoltaics on an accelerated schedule is not imperative.

The United States and the world are facing a watershed decision regarding the nature of future energy sources. Either we shall commit ourselves to further expensive extrapolation along the same old fossil/nuclear track, or we must seek out a new track and begin shifting to it as soon as possible.

What has been lacking is some germinal idea, some reasoned strategy for removing ourselves from the present cul-de-sac.

We have been throwing ourselves at the energy problem, and in the process, working into a societal and psychological morass. It is little wonder, for none of the major energy sources being manipulated—oil, gas, coal, nuclear—will fit into the long range future. Moreover, most of them don't fit the present very well. This society is ripping itself apart as various competing groups tear at each other, trying to build a future from elements that simply do not align with long term needs. It is imperative to look farther down the road.

The Harvard Business School study recognized the problem: "Broadly speaking," it said, "the nation has only two major alternatives for the rest of this century—to import more oil or to accelerate the development of conservation and solar energy. This is the nature of the choice to be made. Conservation and solar energy . . . are much to be preferred."[27]

What is needed is a clearer goal than the muddled, actually derelict, objective of more energy from any and all sources, at staggering economic, social, and environmental costs.

The first order of business is freedom from dependence on supplies of petroleum from abroad. If, by some magic, petroleum use could be cut by 40 percent, this would eliminate all need for foreign oil for a few years at least. The problem with reducing oil use lies with the equipment that depends on it; the automobile alone now consumes just about the amount we import. If there is one culprit, it is the gasoline engine under the hood of the automobile.

---

[27] Stobaugh and Yergin, *Energy Future,* p. 216.

The present fleet of cars used in this country needs to be replaced with cars that don't use gasoline or other petroleum products for fuel. Strangely enough, the machines that have played so large a part in bringing on our present plight could be the instruments of relief.

ELECTRIC AUTOS: A MARRIAGE MADE IN HEAVEN

Until quite recently, any thought of widespread introduction of electric autos would have been amiss, not only because the state-of-the-art in electric vehicles was not sufficiently advanced, but because electric power was frequently in short supply at peak periods. This country is in the process of trying to reduce the use of electric power, and the annual rate of increase has subsided from 5 percent to 2 percent. Nevertheless, use is expected to double by the end of the century.[28] In a situation of short supply, a massive turn to electric power for transportation plainly was not credible. Fortunately, recent progress with photovoltaics now places the whole matter of electric supply in a new perspective.

There is today a remarkable coincidence of events that could be one way out of the energy cul-de-sac just discussed. The key feature in this optimistic scenario is the joining together of two new technologies, photovoltaics and electric vehicles. The circumstances are:

- A widespread realization that fossil fuels are finite, uncertain, and carry high external costs;
- The technical feasibility of massive photovoltaic deployment;
- The development of a new generation of storage batteries that will both lower the costs for electric vehicles and provide greater driving range.
- The fact that the transportation sector of our economy uses just about the same amount of petroleum products as we are currently importing.
- The fact that the total inventory of automobiles is replaced about every ten years.

Photovoltaic cells would not be put on an electric auto, as the available area could not provide nearly enough electricity to propel it. Most homeowners in the future could have enough photovoltaic capacity to power an electric vehicle in addition to meeting normal household requirements.

---

[28] Edward Teller, *Energy from Heaven and Earth* (San Francisco: W.H. Freeman and Co., 1979), p. 292. Also see *Statistical Abstract of the U.S.*, 1979, p. 607: Table No. 1025 shows usage increasing 5 percent per year or more in the period 1980–1990, for a gain of 65 percent for this decade alone.

An electric auto gets 2½–3 miles from 1 kilowatt hour of electricity. Fifteen kilowatt hours, about the amount a typical American household uses each day (excluding space heating and cooling), would propel an electric auto about 40 miles. This is typical urban-suburban daily use of the automobile, amounting to 1,200 miles a month or 14,000–15,000 miles a year. Thus, a photovoltaic array collector of about 325 square feet (30 square meters) would provide enough energy in a year to power an electric auto 15,000 miles. The annual fuel cost for an electric car would be about the same as the annual electric bill for a medium-sized house (without heating or cooling).

If electricity from a utility costs six cents a kilowatt hour (prices in the U.S. range from two to thirteen cents), the price for fuel is less than three cents a mile, compared to perhaps six cents a mile for a compact gasoline car that gets 20 miles to the gallon. Thus, the direct fuel cost for an electric car is roughly one-half that of a gasoline-powered auto.

But electric cars operate on batteries, and batteries cost money. The state-of-the-art in batteries is rapidly advancing, so that by 1982 a set of batteries for a car will last for 15,000 miles before replacement and cost about $1,000 in today's money. This amounts to about seven cents a mile for batteries, which would make the total cost of powering an all-electric auto about ten cents a mile. However the price of gasoline, (and of diesel fuel) will not stay the same, but will continue to rise. Meanwhile, new kinds of batteries under development use less expensive materials than lead and last longer. Since batteries can be made in any size and shape, subsequent replacement with cheaper, more efficient batteries presents no problem. By about 1985, an electric car clearly will be economically competitive with the gasoline version.

Electric vehicles are now in use in many places and have proven their reliability. In England, for example, many delivery trucks operate on electricity. The United States has an active electric vehicle program. There are now about 4,000 of these cars on the road. Close to 1,000 were produced in 1980, although production capacity is much larger than this. DOE projects that some 8–9 million battery-powered electric vehicles will be in use in this country by the end of the century, given normal market growth.

While it is possible to convert a gasoline-powered auto to electric , it is not economically practical to do so. The electric auto will be an entirely new vehicle. A row of batteries will be placed low and on the centerline. This will lower the center of gravity for maximum stability. The DC motor can be located either in the front compartment or over the rear axle. Electric autos will probably have regenerative braking, with a flywheel taking up energy as the car is slowed and releasing it to accelerate again (Fig. 7–1). This greatly reduces the load on the battery and lengthens

Fig. 7–1 Artist's rendering of an electric auto showing motor and gear on left and flywheel energy storage unit on right.

battery lifetime. Friction braking will be employed in a secondary mode. A special form of clutch is required in connection with the flywheel storage unit. The flywheel would be housed in an airtight chamber from which the air is evacuated to reduce friction.

Basically, the electric car is simpler than its gasoline cousin, and it should be more dependable. The electric motor, which will be reversible, has ample torque at all speeds. A simple reduction gear may be employed, but a transmission will not be required. No longer will the owner be concerned about such things as a starter, oil filter, carburetor, automatic choke, spark plugs, radiator, antifreeze, water pump, fuel pump, distributor, muffler and tail pipe, and the crazy-quilt mélange of anti-pollution devices now being superimposed upon this already complex bit of gadgetry.

The electric auto is ideally suited for urban-suburban use. Not only is it pollution-free, but it wastes almost no energy in traffic. When the car is halted in traffic, the motor does not idle, it simply does not run. An electric auto accelerates easily from a stop to any speed within its limits, and operates as efficiently at slow speeds, even at a crawl, as at highway velocities.

In the winter, heating for the passenger compartment is provided by a small alcohol-fueled unit. Cooling in summer is via compressor that operates from the battery.

First generation batteries will be advanced lead-acid (sulfuric) type. They can be repeatedly discharged to 20 percent of full charge (80 percent depth of discharge) and recharged overnight. If the car is driven to work, the battery can be recharged while parked during the day. If it is used for errands, it can be partially recharged during longer stops, if desired. The lead-acid battery will limit the car's range with full charge to perhaps 100 miles, depending somewhat on the load carried and the type of terrain.

New batteries already under development, such as lithium and sodium-sulfur, will within a few years double and even triple the range.

The electric auto would be ideal today as a second car in the typical household, for urban driving in general and for commuting purposes. As the price of gasoline continues to rise, the "E-car" would be used most of the time and the gasoline model would soon become the backup vehicle. Cars under four years of age are driven most of the total mileage today. Thus new electric autos would be used more and, in a short while, would be saving gasoline very much out of proportion to their numbers.

Single-car families also could adjust to the electric auto without great inconvenience in most instances. Battery packs will be readily interchangeable to permit trips of several hundred miles in a day. Alternatively, a gasoline car could be rented for an occasional longer trip, and the owner would still be ahead economically.

The important point is that the cost of operating an electric car will not go up very fast even if electricity costs rise somewhat. It could even decline as more efficient and less expensive batteries, made from raw materials that are cheaper than lead, come on the market. On the other hand, gasoline models face the sure prospect of steadily escalating fuel costs at the whim of foreign suppliers. In combination with photovoltaics, the electric auto should provide a transportation system that is about as inflation-proof as one could hope for. As the efficiency of photovoltaic systems and batteries increases, the unit cost of motive power will steadily decline—a sharp reversal from today's escalating fossil fuel prices.

Photovoltaics and the electric auto would provide a transportation system that is essentially pollution-free. At the power plant end, expanded use of fossil fuels could be curbed and additional nuclear plants would not be needed, while in the cities the largest source of air pollution—gasoline and diesel engines—would be replaced. The latter would provide a positive feedback mechanism: as smog decreased in our cities, more sunlight would come through and photovoltaic systems would become more effective. Working together, the two technologies would be a double-edged sword against pollution.

Fortunately, the incremental cost of converting to electric autos would be minimal. Eighty percent of all cars are scrapped within ten years of their manufacture. The total truck and bus fleet has an average life of only seven years. Trading gasoline cars as they wear out for electric cars is without doubt the most cost-effective way open to free this country from dependence on foreign oil. The net cost would be the R&D expense associated with developing and "proving in" the new species. The U.S. auto industry is now contemplating a major overhaul to cope with the twin problems of the oil shortage and the rising tide of foreign imports. The auto industry is setting out to design the new "world car"—smaller, more fuel efficient, and inevitably more expensive—to recapture its market at home and expand overseas sales. Philip Caldwell, President and Chief Executive Officer of Ford Motor Company, estimates the cost of the conversion from large to small cars for the American industry at $80 billion.[29] "The costs of this transition are unprecedented in American industrial history," comments the *Washington Post* in an editorial. "Of the four big companies, only General Motors clearly has the financial strength to accomplish it unaided. American Motors, now partly owned by Renault, will depend on French resources. Chrysler depends on the U.S. Treasury. Of the four, Ford faces the widest uncertainties . . . ."[30] Since the American public will be paying for this changeover, it is of some concern what kind of new vehicle is to be provided. What no one dares say is that the more fuel efficient gasoline engines will not end our woes. If the average mileage of the auto fleet is doubled within the next six to eight years, an optimistic assumption, the price of oil will also at least have doubled and we will have gained nothing. As the number of cars on the roads increases, fossil fuel usage will not decline, nor will basic pollution problems change significantly.

The time has arrived seriously to consider converting our entire fleet of autos to electric vehicles on an emergency basis. If mandated on a fleet-wide basis, as is being done to increase the mileage of gasoline cars, the changeover could begin as early as 1983 with the fielding of one million vehicles, as indicated in Table 7–2. It would, of course, be necessary to ban the import or sale of non-electric vehicles. By 1984 we could produce and sell only electric autos, with the number of electrics in use increasing each year as shown in the second column (a constant replacement rate of 12 million vehicles a year is assumed.) Reductions in fuel consumption and resulting crude oil savings are reflected in the columns to the right.

---

[29] "Ford Tooling the World Car," *Washington Post,* 21 February 1980.

[30] Editorial, *The Washington Post,* 8 December 1980, p. A-22. Ford, with the assistance of the United Auto Workers, has vigorously urged the International Trade Commission and Congress to restrict Japanese imports to allow the American automobile industry time to catch up. See also, article by Joe Kraft, "Players in the Chrysler Game," *Washington Post,* 9 December 1980, p. A-19. Chrysler was surviving on an unprecedented special government loan of $1½ billion, of which it had used up over half before the end of 1980. Chrysler President Lee Iacocca said, "I have a feeling that we are heading into a wall of solid concrete."

Table 7–2 Effect of Rapid Introduction of Electric Autos on National Gasoline Consumption

Year	*Electric Cars in Use (Millions)*	*Total Miles Operated (Billions)*[a]	*Assumed Average Fuel Efficiency of Old Autos Replaced (mpg)*	*Estimated Reduction in Gasoline Consumption*[b]		*Estimated Crude Oil Saved*[b]
				Gal./yr. (billions)	*Bbl/day (millions)*	*Bbl/day (millions)*
1983	1	15	14	1.07	.07	.09
1984	13	195	15	13.00	.85	1.08
1985	25	375	16	23.44	1.53	1.94
1986	37	555	17	32.65	2.12	2.69
1987	49	735	18	40.80	2.66	3.38
1988	61	915	19	48.16	3.14	3.99
1989	73	1,095	20	54.75	3.57	4.53
1990	85	1,275	21	60.71	3.96	5.03
1991	97	1,455	22	66.14	4.31	5.47[c]

[a] Assumed average annual distance traveled 15,000 miles.

[b] Assumes 1 bbl (42 gal) of crude yields 33 gal of gasoline. Annual rate of imports of foreign oil in summer 1980 was 6.6 mill bbl/day.

[c] 5.47 million bbl/day is equivalent to 11 quads/yr.

The $80 billion that is to be expended to provide smaller gas-saving cars could be used to tool up for the electric auto instead. The buy-in cost of photovoltaics on a national scale would be rapidly offset by the photovoltaic-electric vehicle combination through savings on foreign oil. In 1980 foreign oil imports cost the U.S. $78.6 billion, and costs are sure to go up even though the United States is importing less oil.

Supplies of liquid fuels need to be husbanded carefully to assure adequate amounts for those vehicles that truly require them, such as commercial aviation, defense equipment of many types, the machines for agriculture and the like.

Under normal circumstances, electricity from the grid requires the burning of three times as much energy at the power plant. This is why savants such as Amory Lovins have not been enthusiastic about the electric auto. Under ordinary circumstances, electric autos would not be an appropriate use of energy and would reinforce the trend toward larger central station generating systems. However, if the additional power comes from photovoltaics, the picture changes completely, and the future looks bright indeed!

The two technologies—photovoltaics and electric vehicles—should be brought on in lockstep. At the present state of knowledge, both are entirely feasible, requiring only applied development. By overwhelming good fortune, both photovoltaics and battery technology, the key to electric vehicles, offer ample opportunity for additional breakthroughs.

THE NEED FOR A NATIONAL COMMITMENT

It is important to remember what can be accomplished if there is a genuine national commitment. During World War II, in addition to training and equipping an army and rebuilding a Navy virtually from scratch, Americans set out to build, for the first time, a massive air armada. President Roosevelt set a production goal of 50,000 planes a year. In 1944 alone, the country turned out 95,000 planes. In an atmosphere of near total secrecy, American scientists set out to develop and construct two kinds of atomic weapons with only the flimsiest notion of whether such a thing would work. They went directly from laboratory equipment to full size production plants, skipping entirely the pilot plant stage. In three years time, $2 billion dollars—$4 billion in 1980 values—and a workforce that finally totalled thousands of people were committed to the effort. All this was done when labor was scarce and materials and equipment of every variety were in short supply. The Manhattan Project was given top priority over every other effort in the nation, including armament, manpower, and even military manpower. Both kinds of weapons worked. One type was never tested before use. If the threats to our well-being today are not so immediate, they are no less real.

In the present peacetime setting, we could undertake special projects of seemingly enormous size without noticeable sacrifice of lifestyle. Indeed, because there are so many possibilities and the chance of success is correspondingly high, more than a decade ago it was suggested that a Manhattan-style effort similar to the war-time project that developed the first atomic weapons, be launched to explore the new frontier of photovoltaics.[31] The presumption of helplessness is itself one of the most vexing parts of the whole energy situation. Walt Kelly's "Pogo" in his wise and gentle comic strip had it figured out pretty well: "We is met the enemy," he says, "and he is us."

Photovoltaics seems to be free of the hazards, tradeoffs, and inequities fostered by fossil fuels. Photovoltaics can realistically:

- Ease the balance-of-payments account stemming from overdependence on foreign oil supplies;
- Help reduce pollution caused by the burning of fossil fuels, especially when solar-charged electric vehicles become widely used;
- Curtail long term carbon dioxide pollution of the atmosphere;
- Render unnecessary vast increases in coal mining;
- Ease inflationary pressures stemming from oil price escalation;
- Buy time to resolve production and storage problems of nuclear electric generation;
- Provide a clear signal to the rest of the world of U.S. determination to become energy self-sufficient.

Energy is so basic to this society—energy habits and policies are so deeply embedded in both physical surroundings and social values—that the structure of meaningful change in the energy market must arise from, or at least have the assured support of, a broad spectrum of the populace. In this society energy supplies and basic issues of political and economic health, quality of life, environmental protection, and social equity should be complementary, not competitive, goals.

---

[31] The suggestion is attributed to Karl Boer, University of Delaware, as reported by Professor J.J. Laferski, Brown University, *IEEE Spectrum,* February 1980, p. 31.

FURTHER READING

Abelson, Philip H. "The Oil Price Spiral." *Science* 207, 29 February 1980, p. 937.

Bodansky, David. "Electricity Generation Choices for the Near Term." *Science* 207, 15 February 1980, pp. 721–728.

"Conservation and Solar Energy Programs of the Department of Energy—A Critique." Office of Technology Assessment, U.S. Congress. Washington, D.C., June 1980.

"Convention for a Solar America: Sense of Urgency on Energy Advocated for National Policy." *Solar Engineering*, October 1979, pp. 22–24.

Global Energy Futures and the Carbon Dioxide Problem. Council on Environmental Quality. Washington, D.C.: Government Printing Office, 1981.

Global 2000 Report to the President. Council on Environmental Quality and the U.S. Department of State. Washington, D.C.: Government Printing Office, 1980.

Halacy, D.S. *The Coming of Age of Solar Energy.* New York: Avon Books, 1975.

"Hayden and Lovins Debate Institution Roles." *Solar Engineering*, October 1979, p. 24.

Novick, Sheldon. *The Electric War—The Fight Over Nuclear Power.* San Francisco: Sierra Club Books, 1976.

Okrent, David. "Comment on Societal Risk." *Science* 208, 25 April 1980, pp. 372–375.

The Good News About Energy. U.S. Executive Office of the President, Council on Environmental Quality. Washington, D.C.: Government Printing Office, 1979.

The Electric Vehicle Mini-Guide. 1980. Electric Vehicle Council, 1111 19th Street N.W., Washington, D.C. 20036

[Wayne, Mary]. "The Promise and Puzzle of Electric Vehicles." *EPRI Journal* 4 (November 1979): 7–15.

Whipple, Chris. "The Energy Impacts of Solar Heating." *Science* 208, 18 April 1980, pp. 262–266.

Zaleski, L., and Pierre, C. "Energy Choices for the Next 15 Years: A View from Europe." *Science* 203, 2 March 1979, pp.849–851.

Epilogue

A new technology is unfolding before our eyes—one which could revolutionize our world by augmenting finite energy supplies with limitless energy from the sun. Photovoltaics affords the individual direct access to convenient, versatile, clean electricity. The homeowner can own and control his own energy system.

At the same time, photovoltaics offers unique opportunities for reorienting two other technologies that are central to our civilization—the automobile and the electric utility, both now in dire straits.

It is a dramatic moment when post-industrial civilization is offered a new tool such as this and must choose how to apply it—whether to pick it up and test its possibilities or to cast it aside. The test is whether collectively we have the acumen, the wit, the skill, and daring to put it to good use promptly.

The thought of such change is stupendous and seemingly dwarfs the capability of the individual to have any effect on it. What can the average person do that will help bring about an order that is free from extortion by foreign oil powers, that provides energy without pollution, and that shifts control of energy-producing equipment close to the people who use the energy? While this goal may appear visionary, there are nevertheless concrete actions that citizens can take to assure its realization.

Contacting manufacturers about types and prices of photovoltaic systems for particular needs alerts them to rising public interest. The industry presently sits waiting for clear signs of public responsiveness.

Those who are financially able and willing can purchase and install photovoltaic systems of limited size on their residences. This will help demonstrate the feasibility of photovoltaics and heighten interest in the technology. As small orders come in, industry will be encouraged to expand capacity to meet a growing market. This will maintain the downward trend in prices.

Seriously consider the purchase of an electric auto that can be used frequently for short trips, including commuting to work. In typical urban/suburban use, eighty percent of car trips are less than ten miles long. Purchasing an electric auto will, of course, save gas. More important, it will advance the technology and hasten the day others can follow suit.

Farmers can consider special applications where photovoltaics would work as well as or better than a power line. Electric fences can now be powered by a relatively small photovoltaic array in combination with a battery, and so can a light at a remote location. Such systems seldom need attention. Another frequent application is pumping water from a stream, lake, or well at a point remote from utility power.

Lawyers can investigate local ordinances to see if they restrict the introduction of new equipment such as photovoltaics, and explore ways to remove or ease such impediments. In the spirit of enlightened self-interest, matters such as sun rights, insurability of photovoltaic arrays, and local, state, and federal incentives for solar equipment purchase and installation should be examined.

Insurance companies could prepare now to provide insurance against weather damage, such as hail, to photovoltaic arrays and other coverage that a homeowner or business person may need. Companies that provide coverage for solar systems should mention it in advertisements to the public.

Local government officials can look into the availability of photovoltaic equipment for special applications. Building and zoning ordinances can be modified so that they will not restrict photovoltaic system introduction; new ordinances can be introduced to positively promote photovoltaic uses.

If smog is a problem, tax incentives for people who buy electric autos would be a positive step. One of the most constructive steps that states and localities could take would be to reduce annual license fees and taxes on electric vehicles. Such measures could be justified on the basis that there is a positive contribution to environmental improvement—not an insignificant benefit to the community in some smog-choked urban areas. In addition, special parking spaces could be equipped with metered electrical outlets where electric vehicle owners could get recharges during the day. Many states and localities have already exempted solar equipment, including photovoltaics, from taxation, and others might follow their example.

Local governments could enhance the value of future structures by enacting building codes that encourage, if not require, consideration of maximal south-facing roof areas for future addition of photovoltaic arrays.

Municipally owned electric utilities could agree to buy excess power from cooperating homeowners at reasonable, specified rates. If plant expansion is contemplated, officials should look into the rapidly changing economics of using photovoltaics rather than conventional fuels.

Entrepreneurs might consider opening or participating in a business that installs and maintains photovoltaic systems.

Industries can consider applying solar photovoltaics somewhere in their operations. Special tax breaks might apply, and a photovoltaic array could relieve uncertainties about future prices for oil or other fossil fuels.

Bankers can consider setting up financing guidelines for photovoltaic systems and then publicize them in promotional literature.

Tradesmen can set up a photovoltaic service and installation company, or add this capability to an existing electrical or roofing operation. People with an engineering background may want to look into a photovoltaic distributorship and repair company.

Persons interested in protecting and improving the environment can promote solar photovoltaics, as well as other solar technologies, as a positive step toward reducing air pollution from central power stations; they can call attention to the advantages of electric vehicles for abating pollution in cities and towns.

Chambers of Commerce officials can boost solar businesses as a new area of commercial activity. They can encourage young and growing photovoltaic R&D or manufacturing ventures to locate in their area.

Auto enthusiasts can promote the electric auto through auto hobby clubs. The Electric Vehicle Council* in Washington, D.C. can provide a list of existing electric vehicle clubs. Sponsoring a slide show or a movie on electric vehicles and disseminating literature on the subject can increase public interest.

Advertising executives might suggest to clients some ideas for tie-in advertising with solar photovoltaics. A photovoltaic array in the background of an advertisement provides a look-ahead tone. A photovoltaic array on the roof or in the yard will soon be something of a status symbol, one that also emphasizes frugality and good sense.

Architects can encourage clients to examine the advantages of solar installations and the financial prudence, as well as future convenience, of providing a large clear south-facing roof area at optimum slope for the latitude, and perhaps to plan to use portions of south-facing walls as well. People building new houses or other structures should be advised that photovoltaic systems will be economically competitive in many parts of the United States before the mid-1980s.

---

*A private organization with headquarters at 1111 19th Street, N.W., Washington, D.C.

Building or electric contractors should acquaint themselves with photovoltaic systems layout, installation, and testing, and the benefits when used in conjunction with a heat pump.

Teachers can make photovoltaics a part of the curriculum. High school science teachers can include simple experiments with photovoltaic cells. Students should be given hands-on experience in measuring voltage and current under different lighting conditions. Physics classes could determine "I-V" curves and compute cell efficiency. Biology teachers can call attention to the advantages of photovoltaics and other renewable energy systems from the standpoint of environmental cleanliness. Industrial arts teachers can demonstrate examples of various applications and basic maintenance procedures. College physics instructors can include the principles of photovoltaic conversion and some problems related to basic operational factors. Driver education teachers can include information on electric autos in their courses, and may even be able to arrange a demonstration through a dealer or some individual who owns an electric vehicle.

As Winston Churchill once observed, "In politics the only thing that counts is numbers." Letters and telegrams to congressmen and senators do contribute to the critical decison-making process of government.

Enlightened government policies have nurtured photovoltaic technology along for two decades, bringing it from a laboratory curiosity to the point that it is ready to be used by society. Government can and will continue to help, and in the early stages of commercialization, the government role will be critical. But the federal government cannot force this new technology onto an unwilling or indifferent public, or deliver it on a silver platter.

The political leadership can at best coax and cajole its constituency toward a new energy future. The powerful forces tugging this way and that are too unruly to turn willingly and in unison in a new direction. The politicians will not propel this country into a realm of renewable energy sources. They can't, they won't, and they shouldn't. But when its citizens are ready, the leadership will respond.

APPENDIX 1
A Brief History of Photovoltaics

Until the advent of the space program, photovoltaic technology remained the province of research scientists. Although the conversion of light into electricity had great impact on inquiries into the nature of matter and the universe, valiant entrepreneurial efforts could not move photovoltaics into the marketplace.

The French scientist Becquerel was the first to observe the phenomenon. In 1839, while experimenting with an electrolytic cell made up of two metal electrodes placed in an electricity-conducting solution, he discovered he could generate more current simply by shining light on the cell. Becquerel further found that the amount that the current increased depended on the wavelength of the light.

Willoughby Smith discovered the photoconductivity of selenium in 1873, and Adams and Day observed the photovoltaic effect in solid selenium three years later. In 1883, Fritts described the first selenium photovoltaic cell and, in 1887, Hertz discovered that ultraviolet light altered the lowest voltage capable of causing a spark to jump between two metal electrodes. In 1904, Hallwachs found that combination copper–cuprous-oxide structures were photosensitive.

By 1905, it was known that at any fixed wavelength (frequency), the number of electrons emitted from a substance depended on the intensity of the light bombarding it, and that the maximum kinetic energy (energy of motion) of the electrons varied with the light's wavelength.

In that year, Einstein astounded the world of science by publishing three extraordinary papers. One described the special theory of relativity; a second explained the photo-electric effect. In drawing on contemporary quantum theory which described the discrete packets of energy occurring at the atomic level, Einstein deduced that the behavior of light could be described in terms of separate particles as well as by the more familiar wave metaphor. He theorized that collisions between electrons and particles of light drove the electrons out of a substance; he precisely formulated the maxiumum kinetic energy of the emitted electrons as varying by a universal constant with the frequency of the light, after subtracting the amount of work the electrons exerted to reach and escape the surface of the material. In 1916, Millikan provided experimental proof of the theory.

In 1914, the existence of a barrier layer in photovoltaic devices was noted. Much of the subsequent scientific and engineering work exploited the copper–copper-oxide interaction, but a perfected selenium device drawing on this developmental effort finally outperformed copper by about a factor of ten. By 1941, selenium devices were converting sunlight to electricity with an efficiency of about 1 percent. In the same year, a new technique called a "grown p-n junction" enabled the production of a single crystal silicon device.

By diffusing an impurity through silicon crystals, scientific teams including Chapin, Fuller, Pearson, and Prince[1] at AT&T's Bell Laboratories and Rappaport,[2] Loferski, and Jenny at RCA were able to push solar conversion efficiencies to 6 percent in the mid-1950s. While materials such as gallium arsenide and indium phosphide were found to have higher theoretical efficiencies, silicon still yielded the highest experimental results: factors such as diffusion length and the stability of the material affected final performance. For this reason, and since the extensive prior research in silicon militated against spending large amounts of money to bring the knowledge of alternative materials up to an equal level, silicon continued to be the favored material for use in photovoltaic cells, a position it holds today.

Western Electric began to sell commercial licenses for silicon photovoltaic technology in 1955. Two of the most successful products to emerge were photovoltaic-powered dollar bill change machines and photovoltaic-powered machines which decoded punched cards and tape. Businesses tried to market such devices as photovoltaic-powered hearing aids, radios, highway construction warning flashers, and various toys, but satisfactory technical performance could not overcome the economic advantages of conventional power sources.

The photosensor industry soon diverged from solar power generation. In the camera industry, the use of silicon in light meters lost out to selenium because the latter's spectral response better matched that of the human eye.

---

[1] Dr. Mortimer Prince is acting director of the federal photovoltaic program in the U.S. Department of Energy.

[2] Dr. Paul Rappaport was the first director of the Solar Energy Research Institute, Golden, Colo.

Although researchers had raised the laboratory performance of solar cells to 14 percent efficiency by 1958, market penetration problems appeared insuperable. At one point, RCA, according to Rappaport, came very close to putting the technology back on the shelf—another example of an interesting idea with no place to go.

No one foresaw Sputnik. The successful use of photovoltaic cells in the space program, as described in Chapter 4, provided the impetus for the investigation of terrestrial applications during the past decade.

APPENDIX 2
Major Photovoltaic Energy Projects in the United States

1. The following projects are being procured under program research and development announcements (PRDAs). Construction contracts have been signed for all of these.

Contractor	*Location*	*Application*	*Size (kW_p)*
Concentrators:			
Acurex	Kauai, HI	Hospital	60
Arizona Public Service	Phoenix, AZ	Airport	225
BDM	Albuquerque, NM	Office building	47
E-Systems	Dallas, TX	Airport	27
General Electric	Orlando, FL	Amusement park	110
Flat Panels:			
Lea County Electric Cooperative	Lovington, NM	Shopping center	100
New Mexico State	El Paso, TX	Computer at power station	20
Science Applications	Oklahoma City, OK	Science & art center	150
Solar Power Corp.	Beverly, MA	High school	100

2. Additional major projects under construction or completed are:

Contractor	*Location*	*Size (kW_p)*
Concentrators:		
Mississippi County Community College	Blytheville, AR	250
Northwest Mississippi Junior College	Senatobia, MS	200
Flat Panels:		
Mt. Laguna Air Force Base	Mt. Laguna, CA	60
Natural Bridges National Monument	Utah	100
Papago Indian village	Schuchuli, AZ	3.5
Agricultural test facility	Mead, NE	25
AM radio station	Bryan, OH	25
Georgetown University International Center*	Washington, D.C.	300

Note: In late 1980, 15 utilities were involved on 18 photovoltaic projects, which when completed, will provide over 1,600 kilowatts of generating capacity.
*Planned.

APPENDIX 3

Sources of Additional Information About Photovoltaics

The principal clearing house for all Department of Energy (DOE) reports is DOE Technical Information Center, Oak Ridge, Tennessee 37830. The folder TID-4600 R2 describes the functions and services the Center provides. Individual reports can be ordered directly by specifying title and, if possible, author and report number. TIC publishes bibliographies by subject area. Many users might want *Solar Energy: A Bibliography* (TID-3351), which contains references prior to 1976. In addition, *Solar Energy Update,* which indexes all new additions each month, is available from TIC on subscription. Many public libraries maintain collections of DOE reports on microfiche.

The Solar Energy Information Data Bank, operated by the Solar Energy Research Institute, 1617 Cole Boulevard, Golden, Colorado 80401, periodically publishes a *Reading List on Photovoltaics.* This identifies books and articles by private publishers, as well as government documents considered to be useful source materials on photovoltaics.

The primary source for all government reports is the National Technical Information Service, 5285 Port Royal Road, Springfield, Virginia 22181. Principal publications are available from them at nominal charge. NTIS publishes *Abstract Newsletter on Energy* (NTISUB/097), a weekly guide to new materials in the energy field (subscription price: $80 per year). Also available from NTIS are 150,000 foreign technology titles.

Additional government reports are available from the Superintendent of Documents, U.S. Government Printing Office, Washington, D.C. 20402. The Government Printing Office provides periodic lists of new reports to which one may subscribe free of charge.

Information on federal programs and activities in photovoltaics is available from: Photovoltaic Energy Systems Division, U.S. Department of Energy, Forrestal Building, Washington, D.C. 20585.

APPENDIX **4**

Potential Suppliers of Solar Cells and Arrays

This list is based on available information from the Solar Energy Research Institute. The authors apologize if other U.S. firms which are now shipping photovoltaic cells and arrays have been omitted.

- Applied Solar Energy Corporation
  15751 East Don Julian Road
  City of Industry, CA 91746
  (213) 968-6581
- Arco Solar, Inc.
  20554 Plummer Street
  Chatsworth, CA 91311
  (213) 998-2482
- International Rectifier
  Semiconductor Division
  233 Kansas Street
  El Segundo, CA 90245
  (213) 322-3331
- McGraw-Edison Company
  Power Systems Division
  75 Belmont Avenue
  P.O. Box 28
  Bloomfield, NJ 07003
  (201) 751-3700
- Mobil Tyco-Solar Energy
  16 Hickory Drive
  Waltham, MA 02154
  (617) 890-0909

- Motorola, Inc.
  Solar Energy Department
  5000 East McDowell Road
  Phoenix, AZ 85008
  (213) 244-5459
- Opto Technology, Inc./Solar Systems, Inc.
  1674 South Wolf Road
  Wheeling, IL 60090
  (312) 537-4277
- Photon Power, Inc.
  10767 Gateway West
  El Paso, TX 79935
  (915) 593-2861
- Photowatt, Inc.
  21012 Lassen Street
  Chatsworth, CA 91311
  (213) 882-4100
- SES, Inc.
  Tralee Industrial Park
  Newark, DL 19711
  (302) 731-0990
- Silicon Materials, Inc.
  999 East Arques Avenue
  Sunnyvale, CA 94086
  (408) 737-7100
- Silicon Sensors, Inc.
  Highway 18 East
  Dodgeville, WI 53533
  (608) 935-2707
- Solarex Corporation
  1335 Piccard Drive
  Rockville, MD 20850
  (301) 948-0202
- Solar Power Corporation
  20 Cabot Road
  Woburn, MA 01801
  (617) 935-4600
- Solec International, Inc.
  12533 Chadron Avenue
  Hawthorne, CA 90250
  (213) 970-0065

- Solenergy Corporation
  171 Merrimac Street
  Woburn, MA 01801
  (617) 938-0563
- Sollos, Inc.
  1519 Comstock Avenue
  Los Angeles, CA 90024
  (213) 820-5181
- Spectrolab, Inc.
  12500 Gladstone Avenue
  Sylmar, CA 91342
  (213) 365-4611
- Spire
  Patriots Park
  P.O. Box D
  Bedford, MA 01730
  (617) 275-6000
- Tideland Signal Corp.
  4310 Directors Road
  P.O. Box 52430
  Houston, TX 77052
  (713) 681-6101
- United Energy Corp.
  666 Mapunapuna Street
  Honolulu, HI 96814
  (800) 836-1593
- Xerox Electro Optical Systems
  300 North Halstead Street
  Pasadena, CA 91107
  (213) 351-2351

APPENDIX 5

DOE Price Goals for Photovoltaic Energy Systems

Year	*Major Market*	*Module Price (per peak Watt)*	*System Price (per peak Watt)*	*Price of Power to User (per kWh)*
1982	Remote/International	$2.80	$6–13	$.25–.50
1986	Single Family Residential	.70	1.60–2.20	.07–.11
1986	Selected Intermediate*	.70	1.60–2.60	.07–.11
1990	Investor-Owned Central Station	.15–.40	1.10–1.30	.06–.09

* Intermediate size plants (shopping centers, apartment complexes, industries, small municipally owned utilities, etc.)

APPENDIX 6
Principal Cell Materials and Structures

Thin film materials and cell structures that may be used to meet the long-range cost goal of \$.15–.40/Wp for flat plate collectors.

- **Silicon films**
  Polycrystalline
  Amorphous
- **Non-silicon films**
  Copper sulfide-cadmium sulfide
  Copper sulfide-cadmium zinc sulfide
  Gallium arsenide
  Indium phosphide-cadmium sulfide
  Cadmium indium selenide-cadmium sulfide
  Cadmium telluride
  Cadmium selenide
  Zinc phosphide
  Copper oxide
  Copper ternaries
  Zinc silicon arsenide
  Cadmium silicon arsenide
  Boron arsenide
  Organics
- **Cell structures**
  p-n Junctions
  Heterojunctions

Schottky barriers
Metal Insulator Semiconductor (MIS)
Semiconductor Insulator Semiconductor (SIS)
Luminescent concentrator
Thermal Photovoltaic concentrator
Electrochemical cell

Technological approaches for lowering silicon solar cell costs to meet 1986 goal of $0.70/Wp for flat plate collectors.

- **Low-cost polysilicon ($14/kg)**
- **Silicon crystal growing and wafering**
  Advanced Czochralski
  Casting
  Ribbons
- **Simplified cell structures for:**
  p-n Junctions
  Heterojunctions
  Schottky barriers
  Metal Insulator Semiconductor (MIS) junctions
  Semiconductor Insulator Semiconductor (SIS) junctions

APPENDIX 7

Earth's Energy Balance

This illustration shows the major paths of energy flow into the Earth's system and from the Earth into space. Energy received from the sun must be continuously dissipated into space or the Earth's temperature would rise until it reached a new equilibrium point. Energy escapes into space mainly through reradiation in the infrared portion of the spectrum (upper right in diagram). However an increasing level of carbon dioxide in the atmosphere is gradually closing off this escape route. (General concept and basic values courtesy of M. King Hubert, "The Energy Resources of the Earth," *Scientific American.)*

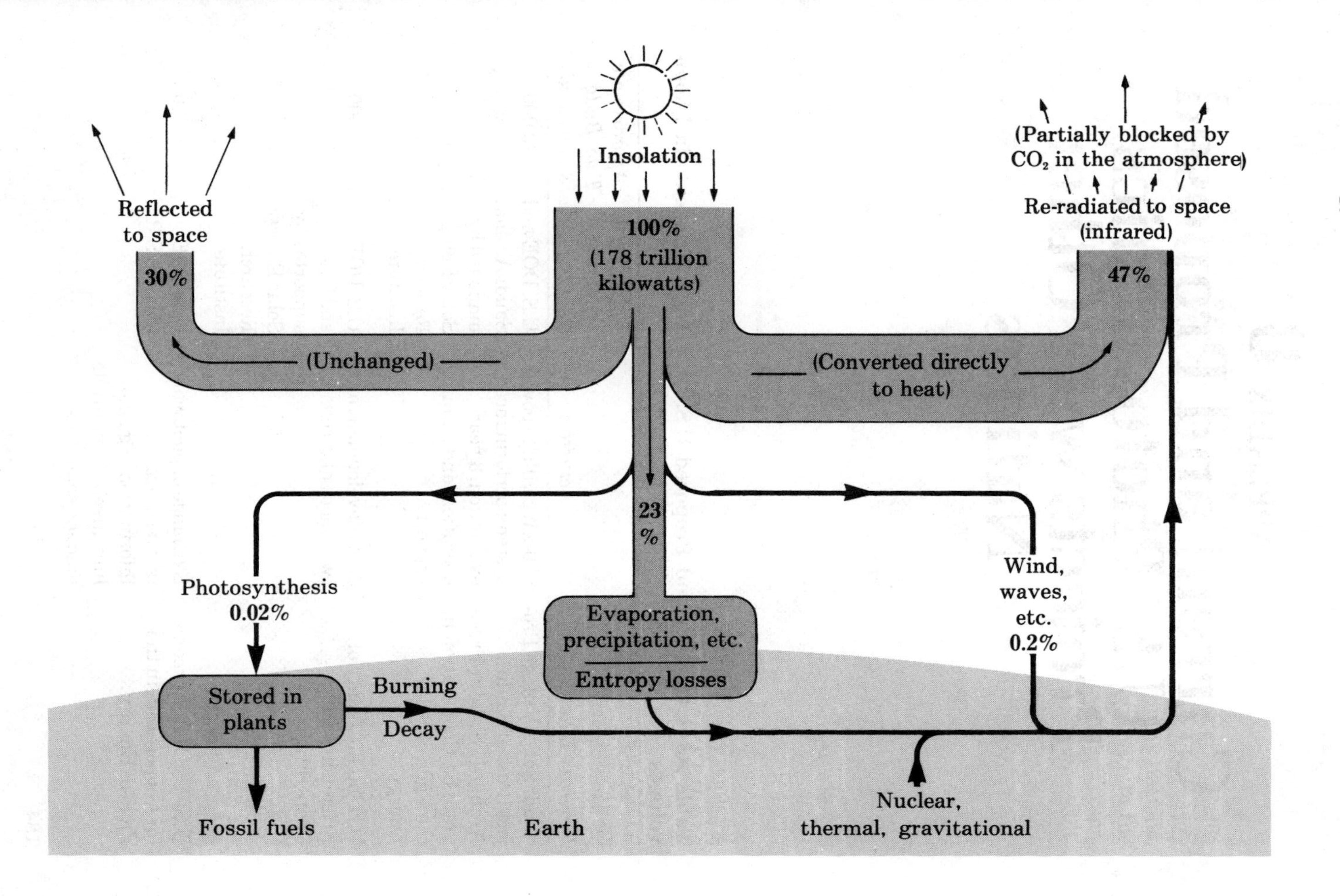
Reflected
to space
30%
(Unchanged)
Insolation
100%
(178 trillion
kilowatts)
23
%
(Converted directly
to heat)
47%
(Partially blocked by
CO_2 in the atmosphere)
Re-radiated to space
(infrared)
Photosynthesis
0.02%
Evaporation,
precipitation, etc.
Entropy losses
Wind,
waves,
etc.
0.2%
Stored in
plants
Burning
Decay
Fossil fuels
Earth
Nuclear,
thermal, gravitational

APPENDIX **8**

Current and Proposed U.S. Photovoltaics Projects with Other Nations

Table A8–1. Current and Proposed U.S. Bilateral Agreements in Photovoltaics

Program Title	*Activities*	*Cognizant Organization*	*Approximate Total Budgets ($ Thousands)*
Saudi Solar Village Project (Part of Saudi Arabia-U.S. Joint Program for Cooperation in the Field of Solar Energy)	500-kWp village power system; performance evaluation, 3 year operation, and training support	U.S. DOE and Saudi Arabia; managed by Solar Energy Research Institute	20,000
U.S.-Spain Treaty of Friendship and Cooperation	Joint development of two-sided solar cells	U.S. DOE and Spain; supported by Solar Energy Research Institute	1,000
U.S.-U.S.S.R. Agreement for Cooperation in the Field of Energy	Exchange of photovoltaic technical information (e.g., gallium arsenide, multiple junction solar cells)	U.S. DOE and U.S.S.R.	30

Program Title	*Activities*	*Cognizant Organization*	*Approximate Total Budgets ($ Thousands)*
U.S.-Italy Memorandum of Understanding for Joint Solar Energy Programs	Proposed 20-kWp photovoltaic experimental system and 3.5–5-kWp remote system, research information exchange	U.S. DOE and Italy; supported by Solar Energy Research Institute	2,000
U.S.-Mexico Bilateral Agreement on Energy	Basic photovoltaic research and village power source demonstration	U.S. DOE and Mexico; supported by Solar Energy Research Institute	Not yet determined
Proposed U.S.-Israel Cooperative Projects on Solar Energy	Proposed research and development on luminescent solar	U.S. DOE and Israel; supported by Solar Energy Research Institute	100
Proposed U.S.-Japan Solar Energy Agreement	Joint research and development in photovoltaics	U.S. DOE and Japan; supported by Solar Energy Research Institute	Not yet determined

Table A8–2. Current and Planned International Photovoltaic Projects at the Agency for International Development (AID)

Project Site and Type	*Cognizant Bureau/Office within AID and Other Agencies*	*Status*
Tunisia		
Photovoltaic power for village. A demonstration village project of photovoltaic cells used with power pumping, lighting, and cottage	Near East Bureau with technical advice from NASA/Lewis Research Center	Planning stage.

Table 8–2 Continued

Project Site and Type	*Cognizant Bureau/Office within AID and Other Agencies*	*Status*
Egypt		
Possible inclusion of photovoltaics in a larger solar/wind project.	Near East Bureau	Planning stage.
Morocco		
Larger renewable energy project which allows for photovoltaic experimentation among other technologies.	Near East Bureau	Advance planning state. Project due by August 1979.
Nepal		
Trickle-charge battery system, for use in radio communication. This system is part of a project in teacher training. Radio receivers and cassette tape recorders in remote villages are photovoltaic-powered.	Asia Bureau	Approved and being implemented. Photovoltaic component is approximately 7–9 per cent of total funding.
Bangladesh		
Radio receiver system, powered by photovoltaic cells, as part of a larger project of flood and cyclone warning for rural villages. Approximately 2,200 radio receivers will be distributed (50 Watts each), most of which will be photovoltaic powered.	Asia Bureau in conjunction with Foreign Office of United States Disaster Assistance	Funds slated for FY80. Photovoltaic component is 20 percent or more of total funding.
Upper Volta		
1.8-KWp photovoltaic-powered pumping system for portable water and grain grinding mill in village. Project is part of larger program on women/development/food production, but also explicitly a test and demonstration of a photovoltaic system.	Africa Bureau	Two year project, implemented March 1979.
Mali		
Pumping powered by photovoltaic cells. Part of larger renewable energy project.	Africa Bureau in coordination with Solar Energy Research Institute; Peace Corps volunteers also involved	Approved and in early implementation state. Five year project.
Tanzania		
Communication (radio) and perhaps refrigeration for health clinics. This is to be a small demonstration project.	Africa Bureau	Planning stage by Tanzania mission

Project Site and Type	*Cognizant Bureau/Office within AID and Other Agencies*	*Status*
Rwanda		
Probably limited photovoltaic component of some kind as part of larger village-level renewable energy project.	Africa Bureau	Planning stage.
Niger		
Demonstration of use of photovoltaics for irrigation pumping as small part of an institution-building project.	African Bureau with Nigerian National Solar Energy Laboratory	Approved and being implemented (early phase).
Thailand		
No photovoltaic installations. As part of a larger energy project, there is an agreement to keep Thailand informed of the state of the art in photovoltaics.	Asia Bureau	Approved July 1979.
Philippines		
The pump irrigation component of a larger nonconventional energy project is photovoltaic powered.	Asia Bureau in coordination with Department of Energy	Approved and being implemented. Photovoltaic component, over a couple of years, is approximately 1 percent of total funding.
Asia General		
Asia conference on renewable energy held in Manila, November 1979, included information on photovoltaics.	Asia Bureau	Approved.
Panama		
Small photovoltaic component of larger energy project. Photovoltaic cells will be used to power radio transceivers and repeaters for remote schools.	Latin America Bureau	Approved and beginning implementation.
Guyana		
Possible use of photovoltaics with health clinic equipment.	Latin America Bureau	Early planning stage.
Coordinated Applications Program		
With research and development and on-site emphasis on photovoltaic powered refrigeration for health clinics. Not tied to specific sites; siting to be done later, in coordination with in-country missions.	Energy Office and Participating Agency Service Agreement (PASA) with NASA/Lewis Research Center	Approved in July 1979.

APPENDIX **9**

Estimating Photovoltaic System Requirements for a Residence

A general approximation of system size can be made as follows: One peak kilowatt (kW_p) of photovoltaic system capacity in an area of moderate sunshine produces about 2,000 kilowatt hours (kWh) of electricity in one year (2,500 kWh in Arizona-New Mexico, 1,600 kWh in New England and the extreme Northwest). This would be produced by a permanently mounted array sloped about 45° in a generally southward direction. For example, a person who lives in a region of moderate sunshine and uses 6,000 kWh in a year (actual amount can be determined from monthly electric bills), would need a 3-kilowatt capacity photovoltaic system (6,000 kWh divided by 2,000 kWh per kilowatt of capacity).

The following simple calculations enable one to estimate more precisely the size of a photovoltaic system required for a residence and the cost of power from it.

1. From your electric bill for a month, note down the number of kilowatt hours used. Divide by 30 to obtain your approximate daily usage (load). Since there is some seasonal variation, it would be more accurate to add the amounts for a year and divide by 365, if you have saved your old bills.

2. On the insolation map, find your approximate location and note the solar energy received daily in that zone. The figures represent approximate kilowatt hours per square meter falling daily on a 45° sloping array, facing southward, as a yearly average.

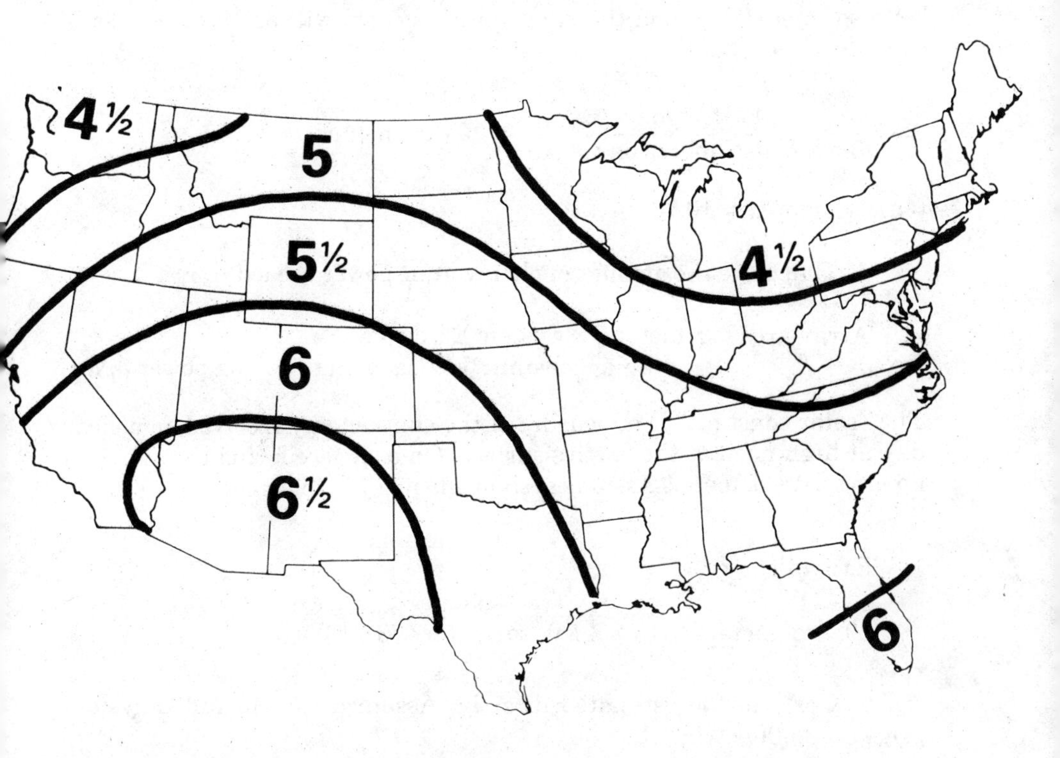

Fig. A9–1 This map indicates the approximate insolation falling in a day on a 45° sloped surface facing generally southward. Figures are in kilowatt hours/sq meter/day (average day in a year). Zones are interpolated from several sources.

(Note: Insolation, or sunshine, is a form of energy, the ability to do work. As with any form of energy, insolation can be measured in any of several terms, e.g., calories, Btus, joules, or kilowatt-hours. The relationships are precise, and the figures can be readily transposed using known ratios. Here the insolation is shown in kilowatt-hours as a convenience for subsequent calculations since the load is given in kilowatt-hours.)

3. Calculate the photovoltaic array area in square meters that you would need to satisfy your average daily electric requirement:

$$\frac{\text{Avg. daily load requirement (kWh/day)}}{\text{Insolation (kWh/sq. meter/day)} \times \text{PV system efficiency}} = \text{Array area (sq. meters)}$$

Example: Assume you use 17 kWh/day, insolation in your area is 5.5 kWh/sq. meter/day, and the photovoltaic system will be 15 percent efficient.

$$\frac{17 \text{ kWh/day}}{5.5 \text{ kWh/sq. meter/day} \times .15} = 20.6 \text{ sq. meters}$$

(One sq. meter = 10.76 sq. ft.)

4. Calculate the amount of photovoltaic power needed:

Array area (sq. meters) × System Efficiency
× Solar constant (kW/sq. meter) = Peak power (kWp)

The "solar constant" is a measure of the sun's energy received on a clear day at high noon at the earth's surface. On a surface facing the sun, it is about 1 kW/sq. meter. Assume system efficiency is 15 percent.

Example:

20.6 sq. meters ×.15 × 1 kW/sq. meter = 3.1 kWp

5. Cost can be estimated directly. Assumed that installed system price is $3,000/kWp:

$3,000/kWp × 3.1 kWp = $9,300

Of the total, 40 percent of the first $10,000, or $4,000, is deductible as a tax credit from federal income taxes; the net installed cost would then be $5,300. (If the system is financed, interest paid on the loan is a deductible item in computing taxable income, like any other interest paid.)

6. To estimate the cost of electricity, assume a system life of 20 years:

$$\frac{\$5{,}300}{17 \text{ kWh/day} \times 365 \text{ days/yr} \times 20 \text{ yrs}} = \$.043/\text{kWh}$$

The final figure for the cost of power does not include any allowance for the "cost of money"—the worth of $5,300 if it had been invested, or the interest paid for borrowing it. This is the largest variable in the estimate.

Tilt Angle (Slope) of Arrays

On a year-around basis, the optimum tilt angle (slope southward) for a fixed array is obtained by adding 10 degrees to the local latitude. In New Orleans at 30°N, optimum slope is 40°; in Minneapolis at 45° N, a 55° slope is best. Exact slope, however, is not critical. For example, a deviation of 15° from optimum reduces efficiency only 3½ percent, and at a 20 degree deviation the loss is only 6 percent (the loss in efficiency increases as the cosine of the angle). For flat plate systems, which pick up light coming from all directions, the optimum tilt angle is less important than for concentrator systems, which must track the sun. If the rotation is on one axis only, the tilt is adjusted manually at least every two or three weeks. Two-axis trackers have sensors so that the array follows the sun continually.

Insolation

On an average day, based on measurements throughout the year, insolation ranges from about 6 kWh/sq meter/day in the Southwest (U.S.) to about 3½ kWh in New England and the extreme Northwest, a difference somewhat less than a factor of two. These values apply to a horizontal surface. When optimally tilted (fixed) arrays are used, the northern latitudes gain proportionally more than the southern latitudes, so that on an annual basis homeowners in Maine in fact have 70 percent of the solar resource at their command as do their counterparts in Arizona, and 55 percent even in winter.

Most areas of the United States receive about one-half the solar energy on a horizontal surface in the winter months that they get in the summer. New England and the Northwest receive about one-third as much in winter as they get in the summer. Again, tilted arrays have a significant moderating influence: the Southwest receives 5 kWh/sq meter/ day in the winter on a 45° sloped surface compared to 6½ kWh/sq meter/ day in the summer, while the Northeast gets about 3 kWh in winter compared to about 5 kWh in summer.

The diffuse component of sunlight is typically about 10 percent of the total in desert regions and about 25 percent of total in urban areas (due to smoke, smog and dust).

Heat to Electricity

If heat is converted 100 percent into electricity, then 3,413 Btu = 1 kWh (theoretical limit). In most conversion processes, about two-thirds of the energy is lost. A convenient rule of thumb is that 10,000 Btu burned under the boiler produces 1 kWh. The conversion efficiency of large steam power plants is about 40 percent. In the case of diesel electric generators, only about 38 percent of the energy in the diesel fuel can be converted finally into electrical energy, the rest being wasted as heat or in other losses.

APPENDIX 10
The Electric Auto

In the early days of the automobile, electric vehicles greatly outnumbered and significantly outperformed cars with internal combustion engines, but well before World War I their popularity slacked and in the 1920s electric vehicles virtually disappeared. The 1970s' oil shortage brought renewed interest in electric vehicles. In 1976, Congress passed the Electric and Hybrid Vehicle Research, Development, and Demonstration Act (Public Law 94-413) which, as amended, directs the Department of Energy (DOE) to develop and put into operation for demonstration purposes 7,500 to 10,000 electric and hybrid vehicles (one that has a small gasoline or diesel engine to extend the range of its electric propulsion system.)

Two prototype electric test vehicles (ETV) were developed and delivered to DOE in 1979 for road testing. General Electric and Chrysler teamed to produce ETV-1 and Garrett Corporation provided ETV-2. Both are 4-passenger subcompacts powered by heavy duty lead-acid batteries. Both have snappy acceleration that approaches present subcompact car performance.

Using the first low-cost, high-power transistor, ETV-1 has: optimized all-electric regenerative braking for extended range; microprocessor-controlled electrical functions; abrasion-resistant plastic glazing; charge indicator replicates fuel gauge function; structure and battery retention designed to ensure crashworthiness; drag reduced through wind-tunnel tests (yaw effects included); integrated electronics (field chopper and on-board charger); and electric interlocks for safe operation and reduced shock hazard.

The first application of composite flywheel technology to passenger cars, ETV-2 has optimized regenerative braking using flywheel energy

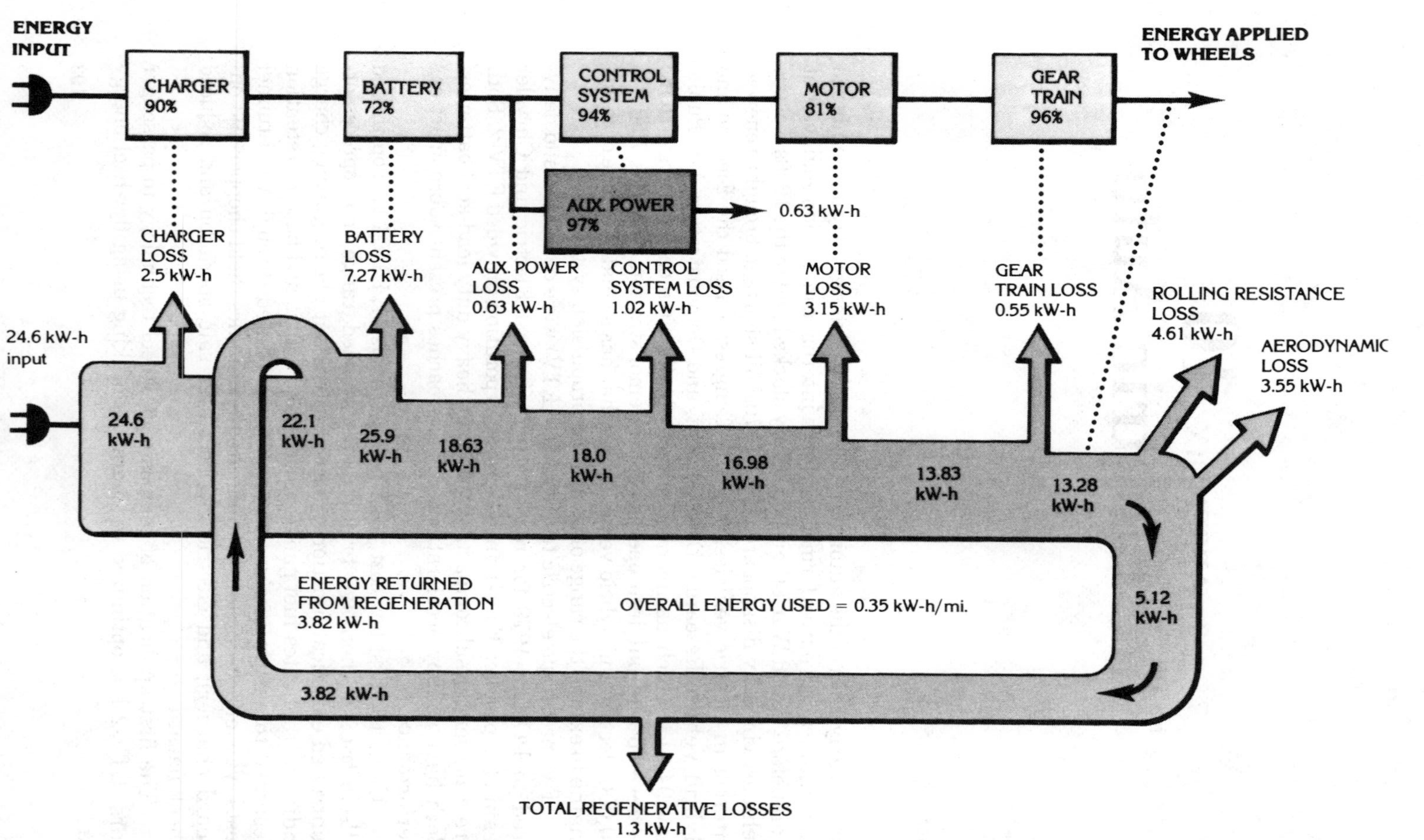

Fig. A10–1 Energy flow model for an electric vehicle. Courtesy of General Electric Co. and U.S. Department of Energy.

storage; flywheel load-levels battery power; microprocessor control; tubular plate applied to golf-cart-size battery; abrasion-resistant plastic glazing; charge and flywheel indicators to replicate fuel gauge function; fiberglass-reinforced plastic in unitized body design; structure and battery retention designed to ensure crashworthiness; electric interlocks for safe operation and reduced shock hazard; and flywheel vacuum maintained by molecular pump.

A typical energy flow diagram for an electric vehicle (Fig. A10–1) shows over half of the energy reaching the wheels, a vast improvement over the internal combustion engine, which wastes at least two-thirds of the fuel within the engine itself, as heat.

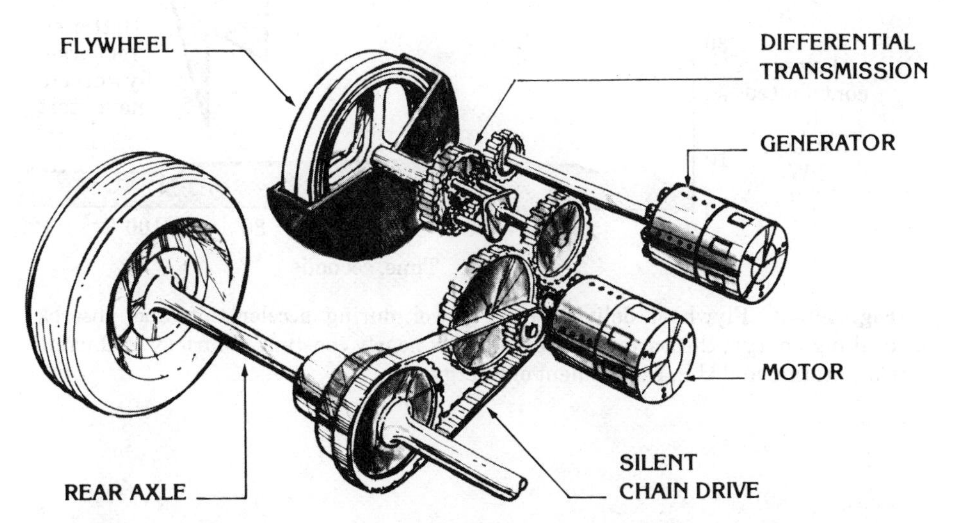

Fig. A10–2 ETV-2 power train uses planetary gearset to control power flow between flywheel, two motor/generators, and rear axle. Courtesy of Garrett Corporation and U.S. Department of Energy.

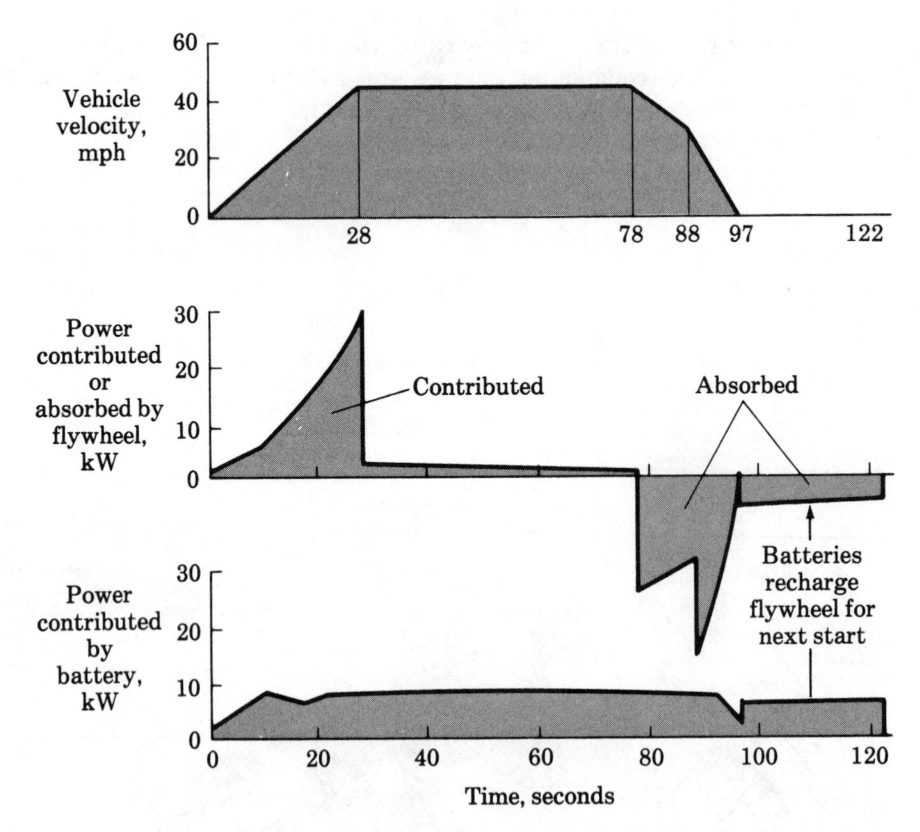

Fig. A10–3 Flywheel delivers peak power during acceleration and absorbs braking energy; the battery power remains nearly constant. Courtesy of Garrett Corporation and U.S. Department of Energy.

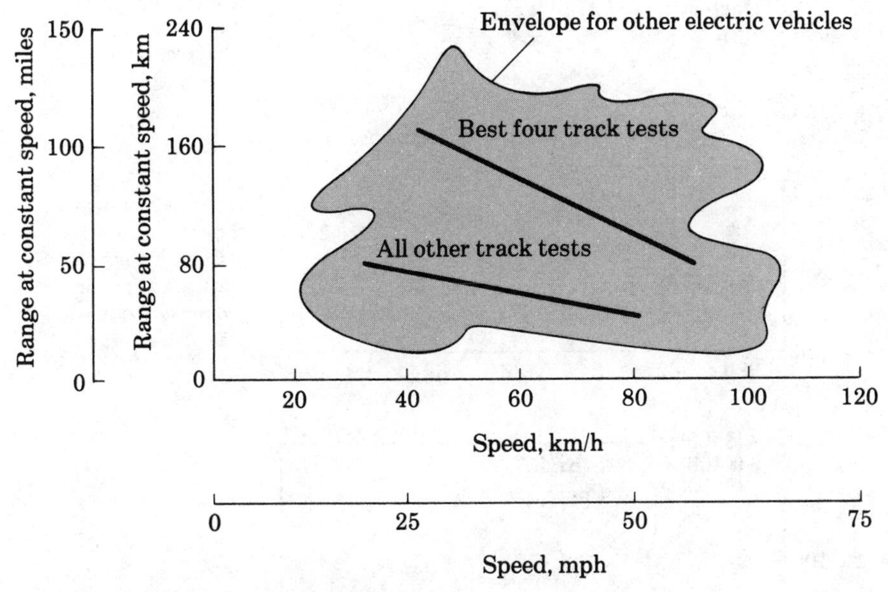

Fig. A10–4 Electric vehicle range as a function of speed. The distance an electric vehicle can travel between recharges falls off linearly with increasing speed. Data from 1976–77. Courtesy of U.S. Department of Energy.

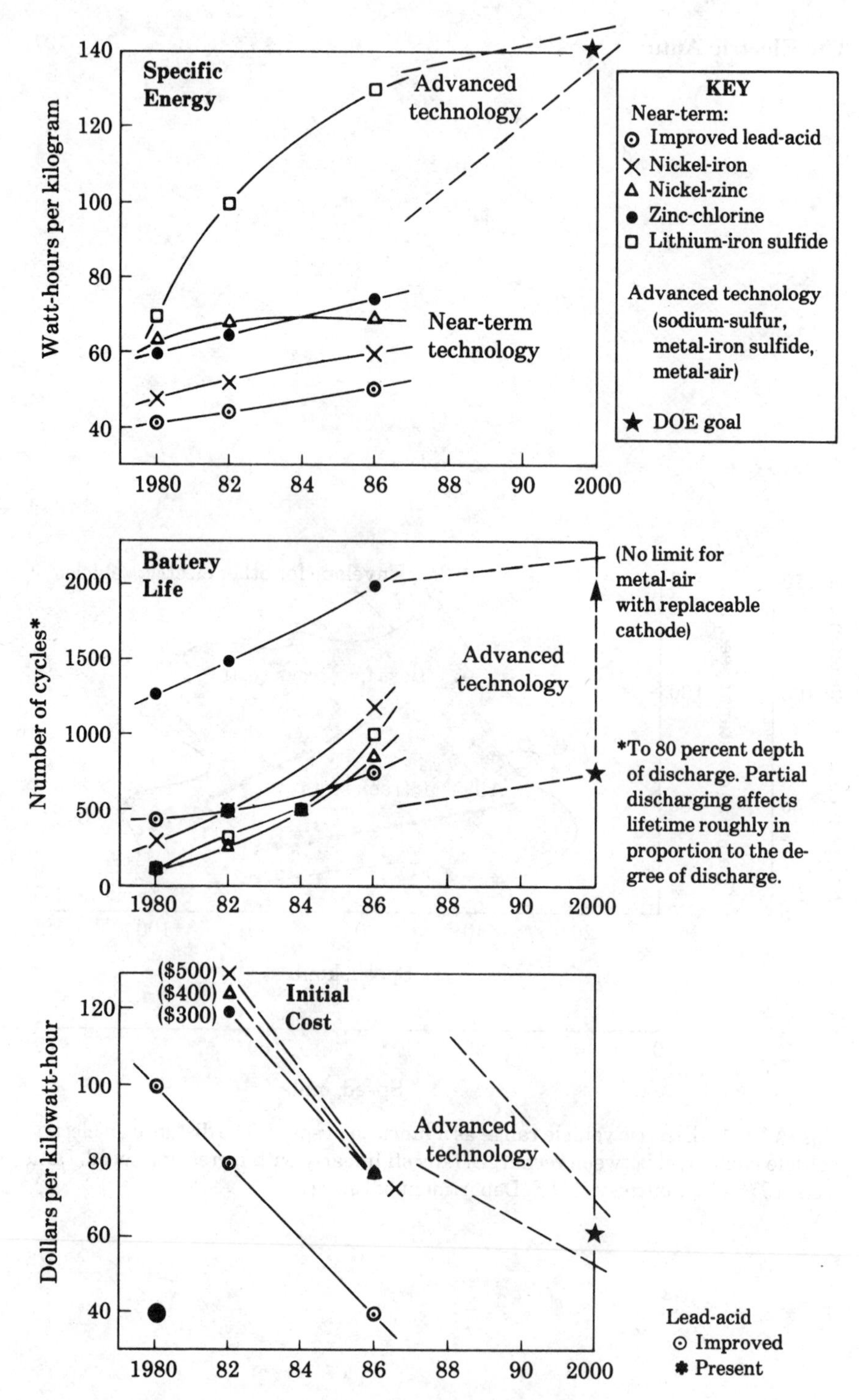

Fig. A10–5 Projected vehicle battery characteristics.

Fig. A10–6 These electric vehicles, developed for the Department of Energy, show the feasibility of using existing automobile technology and components in the near term. They have a range of 40 to 60 miles in stop-and-go traffic. Clockwise from upper left: Converted VW Rabbit produced by South Coast Technology, Santa Barbara, California; Electric van, called Electra Van 600, made by Jet Industries, Inc., Austin, Texas; Converted American Motors Pacer, built by Electric Vehicle Associates, Cleveland, Ohio; Pick-up truck, manufactured by Battronic Truck Corporation, Boyertown, Pennsylvania. Courtesy of U.S. Department of Energy.

FURTHER READING

Assistant Secretary for Conservation and Solar Applications. *Near-Term Electric Vehicle Program.* S-22-192. Washington, D.C.: Department of Energy. 1978.

Assistant Secretary for Conservation and Solar Applications, Office of Transportation Programs. *The Charge of the Future: An Introduction to Electric and Hybrid Vehicles.* DOE/CS-0107. Washington, D.C.: Department of Energy, 1979.

Kirk, Robert S. and David, Philip W. "A View of the Future Potential of Electric and Hybrid Vehicles." U.S. Department of Energy. Paper presented at the 7th Energy Technology Conference and Exposition, Washington, D.C., March 1980.

Klunder, Kurt W. and Katz, Maurice J. "The Department of Energy Electric Vehicle Battery Program." U.S. Department of Energy. Preprint of presentation at the Intersociety Energy Conversion and Engineering Conference, Boston, Massachusetts, August 1979.

"Piecing Together the Electric Vehicle." *EPRI Journal* (thematic issue) 4 (November 1979): 9.

APPENDIX 11
U.S. Energy Sources and Uses: A Comparison Between 1980 and Projections for 1990

Projections for 1990 show dramatic increases in nuclear and coal production to provide more electricity. Unfortunate consequences will be that most of the additional energy generated will be lost in conversion and transmission and pollution problems will be intensified.

U.S. ENERGY SOURCES AND USES IN 1980
(IN MILLIONS OF BARRELS PER DAY OF OIL EQUIVALENT – MMBDOE)

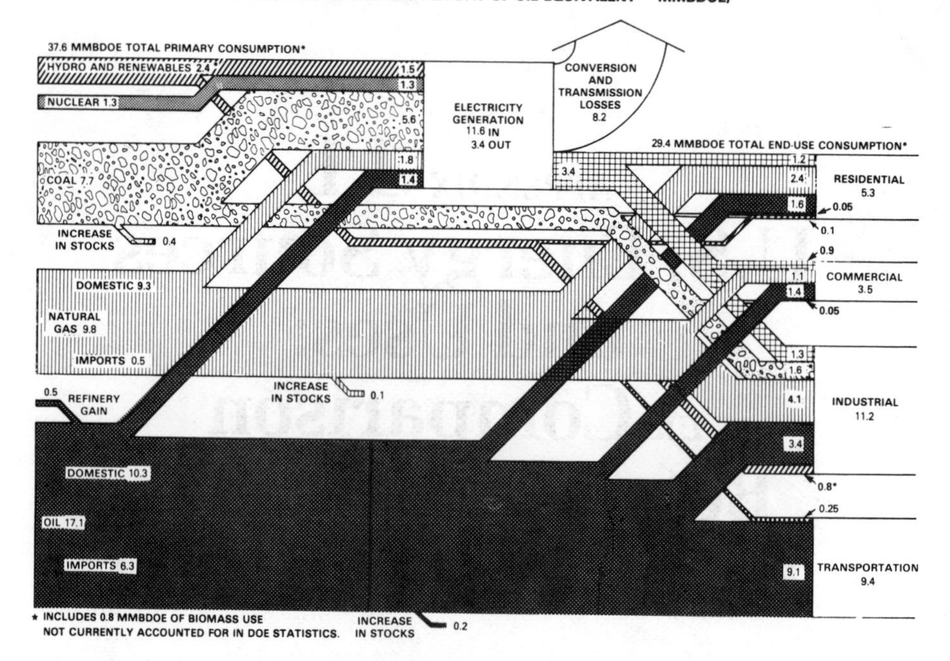

PROJECTED U.S. ENERGY SOURCES AND USES IN 1990
(IN MILLIONS OF BARRELS PER DAY OF OIL EQUIVALENT – MMBDOE)

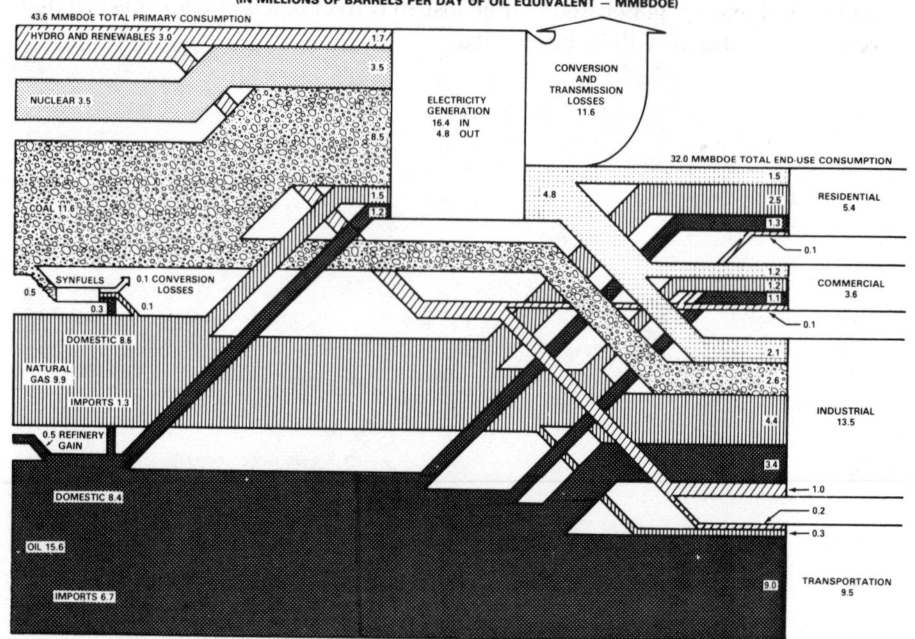

APPENDIX 12
Key Legislation

Solar Photovoltaic Energy Research, Development, and Demonstration Act of 1978 (Public Law 95-590)

This is the principal law, and the only one, that pertains solely to photovoltaics. It prescribes an aggressive research and demonstration effort over the ensuing decade to double production of photovoltaic cells each year, reaching a cumulative production by 1988 of 4 million kilowatts (4 gigawatts), and to reduce the average installed cost to $1 per Watt. This law authorizes spending $1.5 billion to achieve these objectives, subject to annual appropriations. Among other things, the Act prescribes specific studies that must be undertaken, establishes a high level advisory committee, and encourages a strong international effort.

Department of Energy Act of 1978 (Public Law 95-238)

Section 208, *Civilian Applications*, known as the Tsongas amendment, promotes the use of photovoltaic systems at federal installations, and calls for studies of government support for the U.S. photovoltaics industry and of applications in foreign countries.

Crude Oil Windfall Profits Tax Act of 1980 (Public Law 96-223)

One provision qualifies purchasers of solar electric systems, which includes photovoltaics, for a 40 percent tax credit on the first $10,000 of the purchase price (limit $4,000).

Public Utility Regulatory Policy Act of 1978 (Public Law 95-617)

Known as the PURPA, this act is designed, among other things, to encourage the homeowner and other small users to generate part or all of the electric power they require. To this end, it requires that utilities supply supporting or backup power to these "cogenerators," charge reasonable rates for this, and buy back any excess power produced by cogenerators, at reasonable rates based on the "avoided costs" to the utility.

Federal Photovoltaic Utilization Act (Public Law 95-619—Part 4)

This portion of the National Energy Conservation Policy Act, enacted in 1978, was the basis for the federal program to procure photovoltaic systems for government use, one means for building up the U.S. industry and accelerating experience with the new technology.

In addition to this federal legislation, a number of the States have provided tax incentives to encourage the use of solar energy systems, including photovoltaics. The following table provides a summary of state tax incentives that relate to photovoltaics.

A Summary of State Tax Incentives (Spring 1979 data)

	Income Tax Incentive		*Property Tax Incentive*		
State	*Credit*	*Deduction*	*Credit*	*Exemption*	*Refund*
Alabama					
Alaska	●				
Arizona	●			●	
Arkansas		●			
California	●				
Colorado		●			
Connecticut			●		
Delaware	●*				
Florida					
Georgia				●*	
Hawaii	●			●	
Idaho		●			
Illinois				●	
Indiana				●	
Iowa				●	
Kansas	●†				●
Kentucky					
Louisiana				●	
Maine		●			●
Maryland			●	●	
Massachusetts		●†		●	
Michigan	●			●	
Minnesota				●	
Mississippi					

State	Income Tax Incentive		Property Tax Incentive		
	Credit	Deduction	Credit	Exemption	Refund
Missouri					
Montana	●				
Nebraska					
Nevada				●	
New Hampshire				●*	
New Jersey				●	
New Mexico	●				
New York				●	
North Carolina	●			●	
North Dakota	●			●	
Ohio					
Oklahoma	●				
Oregon	●			●	
Pennsylvania					
Rhode Island				●	
South Carolina					
South Dakota				●	
Tennessee				●	
Texas					
Utah					
Vermont	●			●*	
Virginia				●*	
Washington				●	
West Virginia					
Wisconsin	●				
Wyoming					

* In some localities.
† Commercial buildings only.
Source: Solar Energy Research Institute (Printed with permission of SERI).

APPENDIX 13 Miscellaneous Conversion Factors and Approximations

1 bbl. = 42 gal

1 bbl. of crude oil yields 33 gals of gasoline

1 bbl. of crude oil is used in propelling a compact car 1000 miles (at 30 mi/gal)

1 cal. = 4.187 Watt-sec

50 GW_p of photovoltaic capacity replaces in one year 1 Q of fossil fuel consumed to make electricity.

50 GW_p provides in one year 100 billion kWh, or 10^5 GWh, of electricity.

1 horsepower = .75 kilowatts

1 kWh/sq meter = 86 Langleys

1 Langley = 1 cal(gm)/sq cm

1 Langley = .01163 kWh/sq meter

1 megajoule/sq meter = 88.1 Btu/sq ft

1 Q (quad) = 10^{15} Btu

1 Q = 180 million bbls of oil

1 Q = 40 million tons of coal

1 Q = 1 trillion cu ft of natural gas

1 mill bbls oil/day for one year = 2.02 Q

1 sq meter = 10.76 sq ft

1 Watt = 1 joule/sec

Glossary

acceptor—a dopant material such as boron which has fewer outer shell electrons than required in an otherwise balanced crystal structure, providing a hole which can accept a free electron.

alternating current (AC)—Electric current in which the direction of flow is reversed at frequent intervals, 120 times per second (60 cycles per second), as used in commercial grid power in the United States. Opposite of direct current (DC).

amorphous—The condition of a solid in which the atoms are not arranged in an orderly pattern; not crystalline.

ampere, amp—A measure of electric current; the flow of electrons. One amp is 1 coulomb (6.3×10^{1} electrons) passing in one second. One amp is produced by an electric force of 1 volt acting across a resistance of 1 ohm.

array—See *photovoltaic array.*

balance of system (BOS)—Parts of a photovoltaic system other than the array: switches, controls, meters, power conditioning equipment, supporting structure for the array, and storage components, if any. The cost of land is sometimes included when comparing total system costs with the cost of other energy sources.

band gap energy—The amount of energy (in electron volts) required to free an outer shell electron from its orbit about the nucleus to a free state, and thus to promote it from the valence level to the conduction level.

barrier—see *cell barrier.*

barrier energy—The energy given up by an electron in penetrating the cell barrier; a measure of the electrostatic potential of the barrier.

base load—The minimum amount of electric power which a utility must supply in a 24-hour period. Utilities typically operate their most efficient generators (usually their newest and largest) to meet base load demand. See *load, peak load.*

boron—A chemical element, atomic number 5, semi-metallic in nature, used as a dopant to make p-silicon.

break-even cost—The cost of a photovoltaic system (in dollars per kilowatt of generating capacity) at which the cost of the electricity it produces exactly equals the price of electricity from a competing source.

British thermal unit (Btu)—Amount of heat required to raise the temperature of 1 pound of water by 1 degree Fahrenheit.

cadmium—A chemical element, atomic number 48, used in making certain types of solar cells.

capacity factor—The output of a generating plant for a specified period of time, say a year, divided by the output if the plant had operated continuously at full rated capacity for the same period.

cathodic protection—A method of preventing oxidation (rusting) of exposed metal structures such as bridges by imposing between the structure and the ground a small electrical voltage that opposes the flow of electrons, and is greater than the voltage that is present during oxidation.

cell barrier—A very thin region of static electric charge along the interface of the positive and negative layers in a photovoltaic cell. The barrier inhibits the movement of electrons from one layer to the other, so that higher energy electrons from one side diffuse preferentially through it in one direction, creating a current, and thus a voltage across the cell. Also called the depletion zone, or the cell junction.

cell junction—The area of immediate contact between two layers (positive and negative) of a photovoltaic cell. The junction lies at the center of the cell barrier or depletion zone.

central power—The generation of electricity in large power plants with distribution through a network of transmission lines (grid) for sale to a number of users. Opposite of distributed power.

combined collector—A photovoltaic device or module that provides useful heat energy in addition to electricity. See *photovoltaic-thermal system.*

concentrator—A photovoltaic array which includes an optical component such as a lens or focusing mirror to direct incident sunlight onto a solar cell of smaller area.

conduction band; conduction level—Energy level at which electrons are not bound to (orbiting) a specific atomic nucleus, but are free to wander among the atoms.

Conversion efficiency (cell)—The ratio of the electric energy produced by a solar cell (under full sun conditions) to the energy from sunlight incident upon the cell.

current—See *electric current.*

Czochralski process—Method of growing a perfect crystal of large size by slowly lifting a seed crystal from a molten bath of the material under careful conditions of cooling.

deep discharge—Discharging a battery to 20 percent or less of its full charge.

dendrite—A slender threadlike spike of pure crystalline material, such as silicon.

depletion zone—Same as cell barrier. The term derives from the fact that this microscopically thin region is depleted of charge carriers (free electrons and holes).

diffuse insolation—Sunlight received indirectly as a result of scattering due to clouds, fog, haze, dust, or other substances in the atmosphere.

diffusion length—The mean distance a charge carrier (free electron or hole) moves before recombining with another hole or electron. Distances are short, typically several micrometers to a few hundred micrometers. Cell efficiency improves with increasing minority carrier diffusion length.

direct current (DC)—Electric current in which electrons are flowing in one direction only. Opposite of alternating current (AC).

direct insolation—Sunlight falling directly upon a collector. Opposite of diffuse insolation.

distributed power—Generic term for any power supply located near the point where the power is used. Opposite of central power. See *stand-alone, remote site.*

donor—A dopant, such as phosphorus, which supplies an additional electron to an otherwise balanced crystal structure.

dopant—A chemical element added in small amounts to an otherwise pure crystal to modify its electrical properties. An n-dopant introduces more electrons than are required for the perfect structure of the crystal. A p-dopant creates electron vacancies in the crystal structure.

electric circuit—Path followed by electrons from a power source (generator or battery) through an external line, including using devices, and returning through another line to the source.

electric current—A flow of electrons, electricity.

energy payback time—The time required for any energy-producing system or device to produce as much useful energy as was consumed in its manufacture and construction.

EPRI—The Electric Power Research Institute, Palo Alto, California; the research arm of the investor-owned utilities in the United States.

fill factor—The ratio of the maximum power a photovoltaic cell can produce to the theoretical limit if both voltage and current were simultaneously at their maxima. A key characteristic in evaluating cell performance.

flat plate (module or array)—An arrangement of solar cells in which the cells are exposed directly to normal incident sunlight. Opposite of concentrator.

Fresnel lens—An optical device that focuses light like a magnifying glass; concentric rings are faced at slightly different angles so that light falling on any ring is focused to the same point. Fresnel lenses are flat rather than thick in the center, and can be stamped out in a mold.

gallium—A chemical element, atomic number 31, metallic in nature, used in making certain kinds of solar cells.

gigawatt—One billion Watts. One million kilowatts. One thousand megawatts. 10^9 Watts.

grid—Network of transmission lines, substations, distribution lines, and transformers used in central power systems.

heterojunction—Zone of electrical contact between two dissimilar materials. See *homojunction.*

hole—A vacancy where an electron would normally be in a perfect crystalline structure.

homojunction—The zone of contact between the n-layer and the p-layer in a single material, the two layers having been created by doping the basic crystal with other substances. See *heterojunction.*

insolation—Sunlight, direct or diffuse (not to be confused with insulation).

inverter—Device that converts DC to AC.

I-V curve—A graphical presentation of the current versus the voltage from a photovoltaic cell as the load is increased from the short circuit (no load) condition to the open circuit (maximum voltage) condition. The shape of the curve characterizes cell performance.

kilowatt (kW)—1,000 Watts.

kilowatt hour(kWh)—1,000 Watt hours.

load—Electric power being consumed at any given moment. The load that a utility must carry varies greatly with time of day and to some extent with season of the year. Also, in an electrical circuit, any device or appliance that is using power. See *base load; peak load.*

majority carrier—Current carriers (either free electrons or holes) which are in excess in a specific layer of a semiconductor material (electrons in the n-layer, holes in the p-layer) of a cell.

marginal cost—The cost of one additional unit within a group of like units.

megawatt (MW)—One million Watts; 1,000 kilowatts.

minority carrier—Current carriers (either electrons or holes) which are in the minority in a specific layer of semiconductor material. It is the diffusion of minority carriers through the cell barrier that creates a voltage, and constitutes a current, in a photovoltaic device. The process becomes more efficient with increasing minority carrier diffusion length.

multiple junction cell—A photovoltaic cell containing two or more cell barriers, each of which is optimized for a particular portion of the solar spectrum to achieve greater overall efficiency in converting sunlight into electricity. See *vertical multiple junction cell* and *split spectrum cell.*

n-silicon—Silicon containing a minute quantity of impurity, or dopant, such as phosphorus, which causes the crystalline structure to contain more electrons than required to exactly complete the crystal structure. There is no electrical imbalance, however.

ohm—A measure of resistance to the flow of an electric current.

open circuit voltage—The voltage across a photovoltaic cell in sunlight when no current is flowing; the maximum possible voltage.

order of magnitude—A factor of 10; used as a convenience in comparing large numbers.

parallel connection—A method of interconnecting two or more electricity-producing devices, or power-using devices, such that the voltage produced, or required, is not increased, but the current is additive. Opposite of series connection.

peak load, peak demand—The maximum load, or usage, of electrical power occurring in a given period of time, typically a day.

peak Watt or Watt peak—The amount of power a photovoltaic device will produce at noon on a clear day (insolation at 1000 Watts per square meter) when the cell is faced directly toward the sun.

phosphorus—A chemical element, atomic number 15, used as a dopant in making n-silicon.

photoelectrochemical cell—A special kind of photovoltaic cell in which the electricity produced is used immediately within the cell to produce a useful chemical product, such as hydrogen. The product material is continuously withdrawn from the cell for direct use as a fuel or as an ingredient in making other chemicals, or it may be stored and used subsequently.

photon—A particle of light, which acts as an indivisible unit of energy; a quantum or corpuscle of radiant energy moving with the speed of light.

photovoltaic—Pertaining to the direct conversion of light into electricity.

photovoltaic array—An interconnected system of photovoltaic modules that functions as a single electricity-producing unit. The modules are assembled as a discrete structure, with common support or mounting.

photovoltaic cell—A device that converts light directly into electricity. A solar photovoltaic cell, or solar cell, is designed for use in sunlight. All photovoltaic cells produce direct current (DC).

photovoltaic collector—A photovoltaic module or array which receives sunlight and converts it into electricity.

photovoltaic module—A number of photovoltaic cells electrically interconnected and mounted together, usually in a common sealed unit or panel of convenient size for shipping, handling, and assembling into arrays.

photovoltaic system—A complete set of components for converting sunlight into electricity by the photovoltaic process, including array and balance-of-system components.

photovoltaic-thermal (PV/T) system—A photovoltaic system which, in addition to converting sunlight into electricity, collects the residual heat energy and delivers both heat and electricity in usable form. Also called total energy system. See *combined collector*.

polycrystalline silicon; polysilicon—Silicon which has solidified at such a rate that many small crystals (crystallites) were formed. The atoms within a single crystal are symmetrically arrayed, whereas in crystallites they are jumbled together.

power conditioner—The electrical equipment used to convert power from a photovoltaic array into a form suitable for subsequent use, as in supplying a household. Loosely, a collective term for inverter, transformer, voltage regulator, meters, switches, and controls.

p-silicon—Silicon containing a minute quantity of impurity, or dopant, such as boron, which provides insufficient electrons to exactly complete the crystal structure. There is no electrical imbalance, however.

PV—Abbreviation for photovoltaic(s).

quad (Q)—One quadrillion (10^{15}) British thermal units. A commonly used measure of very large quantities of energy. The total consumption of all forms of energy in the United States in 1980 was about 78 quads.

recombination—A free electron being reabsorbed into a hole.

rectifier—A device that converts AC to DC.

remote site—Not connected to a utility grid. See *stand-alone; distributed power.*

reserve capacity—The amount of generating capacity a central power system must maintain to meet peak loads. See *spinning reserve.*

ribbon—A thin sheet of crystalline or polycrystalline material, such as silicon, produced in a continuous process by withdrawal from a molten bath of the parent material.

satellite power system (SPS)—Concept for providing large amounts of electricity for terrestrial use from one or more satellites in geosynchronous earth orbit. A very large array of solar cells on each satellite would provide electricity which would be converted to microwave energy and beamed to a receiving antenna on the ground. There it would be reconverted into electricity and distributed as any other centrally generated power, through a grid.

Schottky barrier—A cell barrier established at the interface between a semiconductor, such as silicon, and a sheet of metal.

semiconductor—Any material which has limited capacity for conducting an electric current. Certain semiconductors, such as silicon, gallium arsenide, and cadmium sulfide, are uniquely suited to the photovoltaic conversion process.

SERI—The Solar Energy Research Institute at Golden (Denver), Colorado. Established by Congress in 1974 (Solar Energy Research, Development, and Demonstration Act) to lead the nation's solar energy research and development program.

series connection—A method of interconnecting devices that generate or use electricity so that the voltage, but not the current, is additive one to the other. Opposite of parallel connection.

short circuit current—The current flowing freely from a photovoltaic cell through an external circuit which has no load or resistance; the maximum current possible.

Siemens process—A commercial method of making purified silicon.

silicon—A chemical element, atomic number 14; semimetallic in nature; dark gray; an excellent semiconductor material. A common constituent of sand and quartz (as the oxide). Crystallizes in face-centered cubic lattice like diamond. See *polycrystalline silicon.*

solar cell—A photovoltaic cell designed specifically for use in converting sunlight into electricity.

solar constant—the strength of sunlight; 1,353 Watts per square meter in space, and about 1,000 Watts per square meter at sea level.

solar thermal electric—Method of producing electricity from solar energy by using focused sunlight to heat a working fluid which in turn drives a turbogenerator.

space charge—Same as cell barrier, depletion zone.

spinning reserve—Utility generating capacity on line and running at low power in excess of actual load.

split spectrum cell—A compound photovoltaic device in which sunlight is first divided into spectral regions by optical means. Each region is then directed to a different photovoltaic cell optimized for converting that portion of the spectrum into electricity. Such a device achieves significantly greater overall conversion of incident sunlight into electricity. See *vertical multiple junction cell.*

stand-alone—An isolated photovoltaic system not connected to a grid; may or may not have storage, but most stand-alone applications require battery or other form of storage. See *remote site.*

synfuel, synthetic fuel—Any of several fuels, usually liquid or gaseous, derived by processing such fossil sources as oil shale, tar sands, and coal.

thermal electric—Electric energy derived from heat energy, usually by heating a working fluid which drives a turbogenerator. See *solar thermal electric.*

thermophotovoltaic cell—A device that concentrates sunlight on a small heat absorber made of metal or other suitable material, heating it to a high temperature. The secondary thermal radiation re-emitted by the absorber is used as the energy source for a photovoltaic cell. The cell is chosen for maximum efficiency at the wavelength of the secondary radiation.

thin film—A layer of semiconductor material, for example polycrystalline silicon or gallium arsenide, typically a few hundredths of an inch or less in thickness, useful in making photovoltaic cells. Use of this material bypasses the costly steps of growing single crystal ingots and sawing them into wafers. Depending on the material, thin films may be produced in different ways, such as withdrawing a ribbon from a molten bath, slowly cooling a molten sheet on a substrate, or by spray coating.

valence state; valence level energy, bound state—Energy content of an electron in orbit about an atomic nucleus.

vertical multiple junction cell—A compound cell made of different semiconductor materials in layers one above the other like a club sandwich. Sunlight entering the top passes through successive cell barriers, each of which converts a separate portion of the spectrum into electricity, thus achieving greater total conversion efficiency of the incident light. Also called a multiple junction cell.

volt, voltage—A measure of the force or "push" given the electrons in an electric circuit; a measure of electric potential. One volt produces one amp of current when acting against a resistance of one ohm.

wafer—A thin sheet of semiconductor material made by mechanically sawing it from a single crystal ingot.

Watt, wattage—A measure of electric power, or amount of work done in a unit of time. One amp of current flowing at a potential of one volt produces one Watt of power.

Watt hour (Wh, Whr)—A quantity of electrical energy (electricity). One Watt hour is consumed when one Watt of power is used for a period of one hour.

Watt peak—Same as peak Watt.

Index